Knights of the Air

Captured German Scout Planes

Knights of the Air

An American Pilot's View
of the Aerial War of the French Squadrons
During the First World War

Bennett A. Molter

LEONAUR

Knights of the Air
An American Pilot's View
of the Aerial War of the French Squadrons
During the First World War
by Bennett A. Molter

First published under the title
Knights of the Air

Leonaur is an imprint of Oakpast Ltd

Copyright in this form © 2011 Oakpast Ltd

ISBN: 978-0-85706-585-8 (hardcover)
ISBN: 978-0-85706-586-5 (softcover)

http://www.leonaur.com

Publisher's Notes

The opinions of the authors represent a view of events in which he
was a participant related from his own perspective,
as such the text is relevant as an historical document.

The views expressed in this book are not necessarily
those of the publisher.

Contents

To My Mother
Whose Love and Faith
Have Been My Strength and Light
I Dedicate this
Little Book

Foreword

It is not my purpose to write a treatise on aviation, nor to delve into the technicalities of the subject—indeed I shall touch them but lightly here and there and that simply for the purpose of explaining some things concerning the air service in France as it applies to the present war. The facts are in no wise secret but, at the same time, they may not be so very well known in America. The science of aerial warfare, though of such recent birth, comparatively speaking, is vast and varied. I could not if I would, and would not if I could, write a text book that would cover even a small portion of the subject. Therefore, the reader may set his mind at rest on that score immediately.

Neither is this a journal or a recital of personal exploits, for I lay claim to no distinction in that respect. Compared with the records of some of the men who are, or have been, pilots in France and whose names someday will emblazon the pages of history, my diary would be as "The short and simple annals of the poor." So I shall endeavour, after this chapter, to use the first person singular pronoun only when necessary in recounting facts which may seem worthy of note and which have come within my observation. For, to be frank, I believe this little book will include observations for the most part and will dwell hardly at all upon experiences.

Time was when the world looked upon an aviator as a superman for he was navigating that mysterious, unstable, and apparently non-buoyant element, air, at unheard of altitudes, in a machine of several hundred pounds dead weight. But that time is almost passed. True, some people still look upon all aviators as something akin to the unnatural and view them askance; even I will admit that there are some fliers who perform feats that seem almost superhuman. Yet not all aviators partake of the nature of demigods, as I well know; for did not I learn to fly and have not thousands of others learned to do the same? Yes, and tens of thousands of others will follow in due time, even as the motor driver has become a matter of course within the past twenty years. The only difference is that some of us entered the game while it was new and comparatively strange.

I may, in all modesty, say that I went to France as an American and volunteered for the air service in 1916, many months before there was serious thought of the United States entering the war. What prompted that resolution? Sympathy for the Allied cause and the things they were fighting for was the principal motive. But underlying that was the boyhood love of adventure and an avid appetite for the tales of chivalry recounting the deeds and exploits of the knights of old. King Arthur and his Knights of the Round Table; Richard the Lion-Hearted; Roland and Ivanhoe—yes, and Robin Hood, too; all contributed their share to my more youthful dreams. But those days of chivalry had passed, the boyhood imaginations had gone, and I was about to settle down to the even tenor of a modern young man's life when this war broke out and I began reading of the exploits of aviators.

Then did imagination again awake. Instead of faring forth clad in shining armour and mounted on a fiery charger I would don fur-lined vestments and ride a steed of metal, wood, and linen, faster than any horse that ever touched hoof to ground. Felt padded leather helmet would replace the one of steel, there would be gloves of wool instead of gauntlets of mail, a machine gun in lieu of a lance, goggles in place of a *visor*.

So, in due time, it came about, after some trouble, expense and persistence on my part I count myself as fortunate, not exceptional, and if I can explain some things having to do with the ways and means of turning an average young fellow into some sort of aviator and perhaps recount some of the facts that I have seen along the way, the purpose of this book will have been served.

Sometimes I think that the shades of the knights of old must arise in envy at the sight of these modern knights-errant faring forth to battle as on the fabled Pegasus, soaring to heights beyond the clouds; not as the vulture seeking the weak and helpless of earth, but as the eagle who breathes the pure air of the higher altitudes and screams defiance to anything that flies; not scattering destruction out of pure wantonness and malice but seeking rather to drive such harpies from the reaches of the air that sustains Justice and Freedom.

Yes, Sir Knights, much of your spirit and your chivalry have been reborn to live again in our Knights of the Air; the foe is not always an honourable one but he must be met, as you met yours, in a manner that leaves no ignoble stain upon the hand that slays. As you were Defenders of the Faith so are they Defenders of Truth, Manhood and Democracy. This crusade carries them not against the ancient Saracen but against the modern Hun. Many of them have already joined your invisible band that hovers among the clouds above that gigantic battlefield, and more will go when, in His good time, it shall be so ordered.

Until that day imbue each one with the courage, the valour and the spirit that were the well-springs of your prowess. They are fighting for the welfare and safety of the human race; for the mothers, daughters, wives and sisters of the world; for the weak and oppressed, for the old and the poor, for the boys and the girls and the babe in its mother's arms. These who cannot fight are not sheltered by castle walls immune to the death-bolts of the enemy but live in open towns, in constant danger of the terror that flies by day and by night seeking what it may destroy, whether it be cathedrals, treasure, art, or human lives.

To what better cause can a man dedicate his efforts, his time and talents and, if need be, his very life, than to give battle to the vandals who know no pity, no mercy, and no law save the destruction of all upon whom they cannot impose their selfish will?

Bennett A. Molter

Pilote-Aviateur
Escadrille N-102
Armée Française

CHAPTER 1

Chivalry Above the Clouds

A hollow square of men, bareheaded and motionless, in horizon blue, stand around an open grave; above, two aeroplanes wheel and bank. Beside the new turned earth lies a long wooden box, at its head a single figure in uniform speaking slowly, solemnly. No hostile plane comes near—if one sails into the gray vault overhead it passes respectfully by.

As the speaker ceases a half dozen men step from the square and the long box is tenderly lowered. With the first motion a long-drawn moan, rising to a wailing shriek, supersedes the dull thudding of the aeroplanes' motors. Both dive headlong toward the grave, shrieking weirdly, symbolizing and almost voicing the grief of the group on the ground. When but a few yards above the grave the wailing ceases, the planes suddenly sweep almost straight upward, as if to go with the soul on the first steps of its long journey. They bank and wheel away, the group on the ground disperses, the grave is filled in. And now, if a Boche plane appears, it is to make trouble.

Thus a French *escadrille* pays the last honours to a fallen comrade. It is only in the air service, even in France, that such a spectacle is possible. The symbolism of the wailing planes and of the escort of the soul of the warrior may be foreign to the more direct mind of either Briton or German. But to the French it is always fitting to dramatize such honour, and so the aviation service can still make a funeral an individual thing, and a ceremony.

The right to remain an individual, whether in life or in death, is the one great distinction that is given to the men in the most dangerous of the military services. Infantryman, artilleryman, engineer, all are swallowed up in the mass while living, and when the end comes it is a death by mass, too, and their very graves—if they have them—are shared with strangers. Even motorcycle corps and submarine raiders are teamed, and no one man stands or falls by himself. Aviation is the only service left where individual effort is at as great a premium as ever and where success and failure both depend on the man himself. The real air fighter, in spite of every sort of squadron flying and team-work, fights and wins or dies—alone.

It is this distinction, probably, that has brought into the air fighting much of the spirit of the knights of old. In many ways the fighting aviators are living much the lives of the heroes of chivalry. Their warfare is that of man to man, as much as were the combats of the armour-clad horsemen in the lists; they live with spectacular death, and they embody beyond all others the spirit of daily, hourly adventure.

It is natural that the chivalric spirit should be strong. Even the Boche, treacherous and brutal in all other fighting, has felt its influence and battles in the air with sportsmanship and fairness. He does not quite maintain the standards of the Allied pilots—of that more later—but he shows unmistakably that he is doing the best that his *kultur* will permit.

Thus air fighting retains whatever is possible, in a life where the one object is to kill, of consideration for the enemy and of courtesy and kindness. There is mutual respect and exchange of civilities, much as there was between opposing knights. This is most frequently shown in the honour accorded to the dead, and also by the fact that no German pilot would disturb the funeral of a French aviator. A remarkable instance came after the burial of Captain Boelke.

When it was learned that the great German Ace had been brought down and was buried close behind the German lines, many French pilots started out with wreaths of flowers to pay honour to their fallen foe. As they crossed the lines the German anti-aircraft guns sent up their usual shrapnel. The planes flew on amid the bursts until above the grave and then, still under fire, began to drop their wreaths.

Immediately the firing ceased, and the pilots went through the same ceremonies as with one of our own dead; they would swoop down to within a few hundred feet of the ground, "wailing" their motors, and then *chandelle*—soar upward—into the clouds. More than a dozen pilots paid this tribute, and all were permitted to return unmolested to their aerodromes. The guns did not open fire again till they had disappeared, but the next time they crossed the lines the battle went on as before.

All enemy airmen receive this distinguished consideration. Men who are killed in other branches of the service are usually hidden in the nearest hole, only the identification tag and valuables being removed to be sent to the government and returned through it to relatives. But with the airmen it is different. When the fallen aviator is a German there is less ceremony than for one of our own men, but he is given a Christian burial and a monument. This monument is always made from the propeller of his machine, the blades being fastened to form a cross, on which are inscribed his name, if known, and the date of his death. Courtesy does not permit the inscribing of the manner of his death, except that he fell in combat.

The same thing is done for an Allied pilot who falls behind the German lines—and most of our men fall there, as the fighting has always been carried into German territory. When the great German retreat was made in the spring of 1917 the advancing Allied troops found many such graves which had been accorded the honours due them.

Air chivalry, too, provides for the carrying of information as to a man's fate to his friends. When a German comes down behind our lines we gather such news about him as we can. If he is dead we learn his name, if possible, and note the location of his grave; if wounded, we find how seriously and to what hospital he has been sent. All this is written down and the note slipped into a little weighted leather bag. A pilot takes this across the lines and drops it over a German aerodrome, with streamers attached so that it will be seen and picked up. The Germans reciprocate this little touch of humanness, and messages drop on our aerodromes from the clouds, telling who of our missing comrades is dead, who a prisoner and who wounded.

With chivalry comes honour, one implies the other. To achieve honour one must possess chivalry, and chivalry without honour is impossible. A well authenticated incident illustrates the strength of this feeling among the aviators of the French flying corps.

When Boelke was at the height of his career, with sixty-three official victories accredited to him by his government, Navarre, the French Ace, challenged him to a duel. Each was the champion of his respective army at that time, though the two had never chanced to meet in combat. Following the etiquette of the days of old, the challenge was written, with the date, hour, and exact location, clearly and carefully specified. The rendezvous indicated by Navarre was over the Forest of Argonne, which was very well known to both French and German aviators and was easy to find and identify from the air. Nav-

GERMAN TWO-MOTOR GOTHA

BUILT FOR BOMBING LONDON, THIS MACHINE WAS BROUGHT DOWN IN BELGIUM

arre described the type of his plane, its individual markings; stated the altitude at which he would fly and wait for Boelke, so there could be no possibility of mistake or misunderstanding. The challenge, with this information, was sewed in a leather bag, Navarre's card being enclosed, and dropped directly on a Boche aerodrome in such manner that the Germans could not help seeing it.

At six-thirty on the evening of the day before the date named in the challenge for the duel, Navarre engaged three enemy planes in combat and was wounded. Great excitement reigned throughout the French aviation service, for Navarre, at that time (1916), was the greatest of all French aces. In their concern over Navarre's injuries, and because of their solicitude for his welfare, the proposed duel was forgotten by everybody—save one. That one was Corporal Pérouse, of Navarre's *escadrille*, who had distinguished himself as a soldier in the *Chasseurs Alpines* but was comparatively new to aviation.

On the morning of the day set for the duel, Pérouse, without making known his plans or purposes to any one, told the chief mechanic of the *escadrille* that he wanted to try out Navarre's plane. He adjusted his belt of ammunition, armed his machine gun and, when all was ready, flew to the field of another *escadrille*, some ten miles away. Here he waited, smoking incessantly all the while, until thirty minutes before the hour specified in the challenge. Then he took to the air and headed for the Forest of Argonne. Remember, he was only a corporal, with less than twenty hours of flight over the enemy's lines to his credit, and he was going to meet the most famous, skilful and dangerous man in the enemy's flying corps. But not only was the honour of his escadrille in question, the honour of France, as he saw it, was at stake; and he was willing, nay anxious, to defend it and if necessary pay for it with his life.

He reached the point and altitude specified in the challenge and waited, circling round and round, dodging bursting shrapnel from enemy anti-aircraft guns. He did not know at what moment or from what quarter the enemy might pounce upon him in an endeavour to gain the first great advantage of surprise; perhaps they might try to ambush him with superior numbers, maybe they would work a trick of some kind to make sure of getting him. He did not know, he did not care; too much was at stake to make these questions of any concern to him. So he waited, still circling round and round, for a full hour—but Boelke did not appear. If he had met Boelke possibly—I say possibly—this tale would never have been told. But as it was there was nothing for him to do but to fly back to his *escadrille*.

During the time he was gone his absence had been noted, as was the fact that Navarre's machine was missing. The *mechanician* related the occurrence of the morning and told of the excuse Pérouse had given for wanting to take Navarre's machine. When the corporal returned to the *escadrille* he was sharply reprimanded for taking Navarre's machine and was compelled to undergo punishment for his breach of discipline. But he took his medicine without a whimper and after it was all over, upon being questioned, as to the reason for his presumption, told where he had gone, why he went, and the result of his trip.

He knew of the challenge, he had not forgotten the time and place; he knew of Navarre's inability to vindicate the challenge and he feared that the honour of the service might suffer in the eyes of the enemy. So he went, facing a possibility that was almost a probability of swift and certain death, but he went for France.

It was from Lieutenant Paul Montariol, of Escadrille F-44, sharing the same aerodrome with Navarre and Corporal Pérouse at the time, from whom these facts were learned, but the story is known throughout France. It should also be known that two weeks after the incident related the name of Corporal Pérouse was on the list of missing. But if he met his death we may be sure that he did so fighting.; displaying that same spirit of chivalry that prompted him to take the place of his wounded and more famous comrade and that he sleeps in honour somewhere back of the German lines.

The famous are not the only heroes; the mail with many victories to his credit has no monopoly on courage, honour and chivalry, though he is entitled to all the homage which is given him. There must be some, no less valiant in purpose and spirit than the greatest, whose lot it is to fall in the first rush of battle, before fame or good fortune smiles upon them; whom the chances of war—call it luck or what you may—destroy before great skill is acquired through the experience of successful combat. The greatest ace was a novice once; perhaps, like "*Some mute inglorious Milton*," the dice of destiny have been thrown against many a brave fellow who, but for some unforeseen and unforeseeable chance that might come at any time to anyone, would have been as great as any. But that is war.

Fame is to the victor; the fruits of success are to the successful; but honour is to all, even unto the least, who valiantly strive amidst the hazards of war, according to their lights and opportunities, to do a man's part in the struggle to make all men free in this great big world of ours. Corporal Pérouse is a type; there are thousands of others like him—French, English, Italian, American and of the other Allied Nations—who navigate the air, giving shot for shot, strength for strength, and blow for blow against the cohorts of the Hun.

CHAPTER 2

The Bird Men

The air service differs from all others in another way—every pilot is a volunteer. No man has ever been conscripted into an aeroplane, and none will be. The spirit of the man must be right, and that spirit cannot be trusted unless it has first shown itself in an effort to enter the service. In the air, survival is to the fittest under the most rigid of tests, and fitness for the air is a thing peculiar and not yet fully understood.

Bravery is essential, daring is taken for granted; Yet this is not enough. Men whose bravery and devotion have been proven again and again in other branches of the service have failed in the air.

This is because of what we call the "reflex"—the power to act instantly and correctly in the face of appalling danger, but without a second for thought. It is a faculty that does not increase with age— rather diminishes. Older men learn to weigh things, they instinctively wait for second thought. In the air there can be no second thought, no considering of pros and cons—only action. So a boy of nineteen usually makes a better pilot than a man of twenty-two, and he better than one of twenty-five.

Delicate boys, even those whose appearance is unrelated in the mind with the endowment of ordinary physical courage, have become the best of pilots. Guynemer, it will be remembered, had been rejected again and again by the army. Among today's stars are men whom the war found almost without individuality, or bordering on effeminacy. And many such have paid the last price of devotion.

Strength is not needed by the aviator, nor quickness of hand. There is no brusque nor brutal action in handling a *chasse* fighting machine. It must be controlled gently—very gently. I remember one of my instructors telling me: "You must handle an aeroplane as you do a woman, very gently. She resents being hurried." A slow and slight movement of the wrist is all that you need to swerve your plane. She responds to the most delicate touch. The Allied fighting planes today are all controlled by the "stick" method; a single handle, like the gear lever of the ball shift type on an automobile, which stands between the pilot's knees, controls everything. A slight backward or forward movement elevates or depresses the machine, a sidewise motion controls the ailerons governing the banking. Some of the American machines and the Allies' bombing planes have the "Depp" control, a wheel, like the steering wheel of an auto, which gives elevation control by being pulled or pushed to or from the pilot and banks the machine when twisted. But they are alike in this—an inch of movement is more than enough, and it must be made slowly and gently.

It is the spirit of adventure, the lure of danger and the love of excitement that brings most of the volunteers into the air service. Every man in it has shown the steel in his nature somehow, but they are men of the most varying types and experiences. Look at the list of the Americans flying for France:

One was editor of a syndicate of newspapers before he took to the air. One was a dancing star. There was a millionaire whose escapades had furnished much copy to the newspapers, but who had done little else. One man is an author—fairly well known. One entered the service of France within twenty-eight days after he received his honourable discharge from the United States navy. There are two racing auto drivers and three lawyers. There are boys who have come straight from school; some direct, but more by way of the Foreign Legion. There are many of these, boys who have been case hardened by their experience in the terrible attacks made by this world-famed force. The *Legion Etranger* is the great attacking regiment of France. Its regimental colours carry more citations and decorations than those of any other body of soldiers, and those who have been graduated from it are men.

The class of men in the French aviation service, to look beyond our own boys, is the very highest. In my own *escadrille* there was James de Rothschild, son of the multi-millionaire financier of Europe. There was the Viscount de Procontal, son of the marquis whose family is of the oldest French nobility, dating back to before 900 A. D. There were men celebrated in other ways—Dupre, the creator of the game "*Diabolo*," who gave exhibitions to the late King Edward VII, and occasionally amused us in our off hours at the hangars, chasing and catching the elusive spool. There was an artist whose paintings had been hung in the salon of the Grande Palais, and so on. Of course, there were famous fliers—Batcheler, an old time pilot, who flew with Garros in Egypt in the infancy of aviation, and Rene Simone, who astonished America in 1911 by his "wild" stunts in a Bleriot monoplane. They were hardly the A B C of what he does today.

I remarked that daring is taken for granted in a pilot. But there must be more than that. The pilot must act as if everything were all right, even when he *knows* it is not. He must learn to disregard the hazards of his work and to act—always, always—as if they did not exist. He must not only fight as if he expected to win every battle—he must actually believe that he is safe and that Fate has delivered his adversary over to him. Confidence, complete, colossal—that is the keynote of success in fighting pilots.

With this confidence goes something akin to superstition, which is a real strength to many. It would be folly to deny that it has even contributed to more than a few miraculous escapes or victories. Aviators are superstitious. Many have pictures, charms, or religious medals fixed on the dash of their machines as a protection against evil. And nearly every aviator in France wears, chained to his wrist with welded chains, little holy medals given by his mother, wife or sweetheart, without which he would not dare to fly. These give him confidence and keep him fearless, flaunting death daily, week in and week out, till the Great Adventure claims him.

The feeling still persists everywhere that, even apart from fighting, flying is terribly dangerous. In reality it is almost safe. Of course, there is danger in every mechanical contrivance of a mechanical breakdown, and in aeroplanes motor trouble or some weakness may develop. But, if there are no guns below, these can usually be remedied very simply by landing at the first sign of trouble.

Sometimes, of course, it may be difficult to find a safe and smooth landing place; but at the front, if one has a fair height, he can always manage to glide to one of the aerodromes. Even in the fighting planes, which, with their small wing spread, depend to the greatest possible extent on their engine power to keep them up, a good pilot can glide four feet for each foot that he drops—that is, if he is a mile high he can travel four miles to find a landing place. And the direction of the wind is also a factor.

There is safety in height, too, in more than one way, especially for the fighting pilot. In a combat at, say, 17,000 feet a pilot may fall

10,000 feet with his plane completely out of control and still make his escape. If he loses consciousness from a wound during the fight the value of height is still greater, for he has a chance to revive during the downward rush and to save himself. The terrific speed of his fall, the changes of atmosphere, the sudden pressure on his ear drums, the effect on the heart, and the rush of wind all help to revive him. The higher he is when the fall starts, the better his chances of regaining consciousness and control of his machine before he strikes.

Even in bad landings and in short falls most men are saved by the construction of the present day machines, which break the force of the fall and collapse around the pilot, leaving him little the worse. Especially is this true of the *chasse* machine—the fighter. In the older type, where the motor and the propeller were behind, if the plane struck nose first the weight of the motor would often bring it down on the pilot, crushing him to earth.

But the modern *chasse* plane is a tractor, with the motor in front, and as it is the centre of gravity in the machine it strikes first and absorbs most of the shock. The wings may be torn off, the fuselage may crash to splinters, even the *nacelle* may collapse, and yet the pilot, strapped to his seat, may be only cut or shaken up. I have known of this happening time and again.

Another danger that is greatly over-estimated is that from anti-aircraft guns. Dodging shrapnel is not very pleasant, but very few machines or pilots are lost by this cause. Riddling the plane with shrapnel balls makes little difference, so long as the pilot, his propeller, his controls, his engine or his gas tank are not hit. Machines seldom come in without a few holes in the wings—often half the wires and braces will be scratched, and still the landing will be safely made. So common is this shooting-up of machines that it is taken for granted that no machine can go out again till it has been overhauled and repaired. Yet the pilots suffer little from the shrapnel.

Flying among bursting shells, however, has a terribly nerve wracking effect which wears many good men down in time—a sort of aerial shellshock. When this comes, only a long rest away from the front will bring the pilot back to a condition where he can safely go aloft again. The air service is no place for a nervous man. Remember that in piloting a scout machine you have your motor, your gas pressure, your thermometer, your direction, your piloting, your plane and the shrapnel to watch—*as well as the Hun*. The main thing is never to lose your head. You *must* be cool and collected at all times, never allowing an instant for any emotion.

But an aviator's life is not a continual tangle of danger; he has many pleasant hours while he is on *terra firma*. When he is not in the air he lives in far greater comfort than do men in the other services. His means of transportation are so good and so rapid that his cantonment is usually out of range of all but the biggest guns, and in danger only from cruising enemy pilots. The mud is nothing in his life. He can keep clean, and since flying must usually be in good weather he suffers little from the elements. He works in benumbing cold, of course, but his heavy clothing protects him greatly and the warmth from the engine, always ahead of him in the fighting planes, helps much. His quarters are quite comfortable, except for lack of fuel in sharp weather.

Most *escadrilles* at the front are housed in cantonments of portable buildings about 30 feet wide and 80 feet long. These are divided into rooms 12 by 15 feet, each containing two beds and the equipment of the two pilots who share it. At the end of the building is a room 20 by 30 feet, used as an assembly room, with one large table to seat the entire *escadrille* at mess. A mess fund, amounting to about $1 a day, provides, in addition to the staple army rations, many little luxuries. The food is good; better in fact than most of the civilians hundreds of miles from the front enjoy. There are no meatless days, but frequent wheatless ones. However, the black bread is not bad, once you become accustomed to it. France feeds her soldiers well, and well she may, for there is no place where it is more necessary to have every man at top efficiency all the while. Every ounce of food saved here in America or elsewhere helps increase the fighting power of the men Who must win the war.

The aviators have their full share of all the amusements which go on behind the front, and, since they are back of the lines every night, they get more of these than do the "mud-sloggers." In a village, a mile from one of the aerodromes, a detachment of the British Army Service Corps established a movie theatre, putting it in an old barn right next to a gun emplacement. The program was changed twice a week, and it was especially pleasing to me, as most of the films were American ones. It was always all-comedy, the lieutenant in charge explaining that his men must be cheered, and so we were fed with no problem plays nor tragedies which might add even for a moment to our own mental burdens. We watched Charlie Chaplin and laughed while shells fell about us and the big gun next door was booming away like clockwork.

The audience was an Allied one—British, Belgian, French and American. Officers and privates could all understand Charlie Chaplin's language easily enough. A charge of 6 cents for privates and 20 cents for officers was made, and the surplus, after operating expenses were deducted, was used to purchase more machines and films for other sections of the front.

On rainy days, or during snowstorms in winter, when there can be no flying we spend our time writing letters, playing cards, overhauling our kits and tramping. There is always a little interest in the efforts of the Boche long range guns to land a shell among us. Often we ramble up to the front line trenches for a bit of excitement and to see how the war is going on. From our great height in the air we know almost nothing of what is doing in the trenches, and we are almost as interested as a civilian would be in getting a chance to go along the firing lines and talk with the infantrymen.

The one real danger to life in the cantonments comes at night from Boche aeroplanes. They are often most annoying, with their careless dropping of bombs around us. This always starts the anti-aircraft guns going, and then for an hour or so our sleep will be broken. The Boche has learned, too, that he has a much better chance of doing damage if he uses gas bombs, which spread danger over a wide area, than if he sticks to explosives, so we have to sleep with gas masks at our pillows and hustle into them at the first alarm.

I have mentioned that the German does not follow all the rules of fairness in the air game. His chief offense has been in disguising his machines. The air service has always held to the naval law—to use disguises if desirable when cruising around, but never to go into action under false colours. The Boche breaks this rule constantly. At first he used to paint the Allied red, white and blue bull's-eye on his planes, to deceive the Allied gunners on the ground, but for a long time he would put a little Boche cross in the bull's-eye, so that the disguise was not quite complete. Now, however, he leaves even this mark off and tries to appear as a friend up to the time when he opens fire with his machine gun.

All in all, the air service is the one which should appeal most strongly to the American boys and one in which they should excel. Our experience in athletics and sportsmanship, our trained ability to act instantly in all the emergencies of the games which we play, our optimism and daring, should make our boys, with a little practice, the peers of any airmen alive. The records of the few Americans who have flown so far all point that way.

The aviation section is really the *corps d'elite* of the French army, and the members of it have won their places by merit. The soldiers and officers assigned to it have gained that assignment by some gallantry in fighting on the earth. The French fliers who have not served an honourable apprenticeship in the trenches are those who for physical disability or for some other reason are exempt from military duty and have volunteered their services, regardless of their exemption.

The French believe that the fact that they offer themselves for service entitles them to places in the *corps d'elite*, if they show themselves fit for it. Captain Guynemer—the Ace—had been rejected for military service and was a volunteer. Not the least compliment that France has paid to America is that she considered those Americans who offered themselves for the air service also entitled to the great reward.

In picking the men for the scouting and patrol work, however, the severest tests are applied. They are the ones who must do the fighting in combats where the loser whirls two or three miles to death; they are the ones who must know enough to refrain from fighting when their mission forbids it.

CHAPTER 3

Aviators in the Making

There are three distinct varieties or grades of training schools for aviators in France, in which the different types of pilots are trained and instructed for their prospective work. When the need arises for pilots of a certain type or to do a certain work they are selected from the school giving the sort of training necessary to fit them for that particular duty.

If the fledgling pilot is to fly a scout or pursuit plane he is trained for that work and that alone. If he is to fly a photographic or artillery regulating plane, he is trained in another school on a different type of plane. If his work is to be in a bombing plane his training takes place in a third kind of school.

This selection is now done by a process of elimination, which I will touch upon later. But this system was not always in vogue, especially in the early stages of the war. At that time the number of planes was relatively small, as compared with the present time, and everything concerning aerial warfare was in a more or less experimental stage. Therefore pilots were trained to fly whatever types of machine the French government had or could get. It was simply a case of using to the best advantage such equipment as was on hand and so it was necessary *to fit the man to the machine*. Of course the different branches of the service had not been developed and segregated as they are today. But this has all been changed now so that the endeavour at the present time is to build *machines to fit the men* who are trained and ready for service as pilots of the several types.

The making of a *chasse*, or pursuit, pilot depends, mainly, upon the man's physical and mental equipment, temperament and adaptability. These qualities determined, it is then largely a matter of training by means of theoretical and practical instruction and actual practice. By separating the men into groups or classes, giving them certain tests and standards of efficiency to attain, the characteristics and talents of each individual are brought out and observed. In this manner the ones that seem best adapted for this or that branch of the service are segregated and are then instructed and trained along certain well developed lines in the work and duties peculiar to the branch to which they are assigned.

For the *chasse* pilot, whose work is probably the most spectacular and therefore the best known, the best preliminary training, undoubtedly, is that of the Bleriot monoplane. At least that is the French theory and as I took my training in the French schools it is perhaps only natural that I should be partial to that theory. The theory and practice in the British and American schools differ somewhat from those pursued in France.

While the Bleriot training is not the quickest it is the best and, incidentally, the most trying and expensive. It probably involves the breaking and smashing of more machines, and more bumps for the student pilots, but it gets results, one way or the other, and a man ambitious to gain skill and fame as a pilot must not expect a soft and easy time during his training. The French government bears the expense and is willing to do so, if it gets the right kind of pilots.

The principal function of the Bleriot training is to develop the student's "reflex," which is so necessary in flying the high powered machines under war time conditions. It is hard to define all that is included in this use of the word "reflex." It is a sort of subconscious ability to do the right thing at the right time, surely and instantly, without waiting an appreciable length of time and without the intervention of thought or reason. It is an intuitive faculty by which our bodily movements are controlled by instinct rather than by a conscious functioning of the brain.

For example, if a missile is thrown at our heads we may dodge it before the brain records the fact that danger is coming and before reason tells us that we must move the head in order to avoid danger. If an insect flies at our eyes we wink before we are conscious of seeing it. A skilful boxer ducks or blocks a blow from, his adversary, and perhaps whips back a punch at the other fellow, without stopping to analyze the situation or to theorize on the facts involved and the proper answer to them. These are examples of reflex and it is this faculty that must be developed to the highest degree in order to make a successful pilot. The machine becomes almost a part of the finished pilot; he guides it, dodging, twisting and turning, as a bird flies, without conscious effort, leaving his mind and senses free to grasp and record facts and conditions which require the use of judgment so that reason may shape his larger course of action and determine his future movements.

It is the French theory that this reflex should be developed *from the very beginning* of an aviator's training and hence the reason for the preliminary training on the Bleriot single control monoplane.

In these schools an aviator learns to fly very much as a baby learns to walk, a step at a time, holding on to some support at first and receiving the benefit of advice and encouragement from his seniors, but, after all, depending on his own legs and strength, so far as actual locomotion is concerned, from the very beginning. Or, to change the simile, it is like a boy learning to swim; someone tells him how to use his legs and arms and at first he goes into shallow water, near the shore, and learns to paddle and keep himself afloat, but without any actual physical support from his instructor. Later he ventures out a little farther into deeper water, gradually increasing his range and improving his stroke as he gains in strength and self-confidence. Then he learns to dive, to float, to swim on his back and to do many other tricks. The water loses all terror for him; its depth is no longer of any concern, he knows how it acts and what to expect under certain conditions; he has learned how to handle himself in it He has become an expert swimmer not only by following instructions but by relying all the time upon his own wit, intelligence, strength and increasing experience—in a word, he has mastered an unfamiliar element and makes it serve him, instead of destroying him.

Air, as a buoyant element, is unfamiliar to most people and travel through it has its terrors for the great majority simply because it is strange and outside the range of their experience. Height frightens them because experience has taught them that increased altitude means added danger, under ordinary circumstances. But height to the aviator is like sea room to the mariner; the more he has of it, barring a serious accident which throws his machine out of all control and hope of recovery, the safer he is to manoeuvre at will and to select a convenient and safe landing place. In the event of ordinary accident, he is, generally speaking, safer at a good height than near the ground, because he may have time to correct the trouble while he is in the air and to bring his machine back to control—and a few seconds are sometimes sufficient to do this—whereas he would not have time to do so if he were too near the ground. The old adage holds very true in aviation; it isn't the falling that hurts, it is the sudden stop that does the damage. Anyway, if the aviator does fall out of all control it may as well be from 5,000 as from 500 feet, so far as the final result is concerned, just as a man may as well be drowned in seven feet of water as in seventy. In the French aviation schools a man learns to do by doing—from the very beginning. He is taught and *made* to depend upon himself and his own resources from the first day.

His very first ride in a plane is by himself, all alone. His teacher tells him what to do and how to do it, of course, but all of that instruction is received on the ground while the machine is stationary. The instructor does not ride with him on his first trip or at any other time. In the first place, the planes are all single seaters and there is no room for the instructor; in the second place, it would be contrary to French precedents. The student is turned loose to make or break. If he makes good, all is well and he goes on; if he breaks he gets one more chance and that is all. The instructors make their observations of the student's performance from the ground, while the student is in action. After each succeeding test and trial is performed the instructor makes his criticisms and corrections and then the student receives additional information in preparation for the next step, progressing as fast as his ability and accomplishments warrant.

The student is initiated into the mysteries and vagaries of aeroplanes by being strapped into the seat of a *penguin*, a machine resembling the bird from which it derives its name insomuch as, though having wings, it cannot fly. Its wing area is so small, in proportion to its weight, that it is impossible to get it into the air; it can only roll along the ground. It has a ground speed of about thirty miles an hour.

In the *penguin* the student learns the use of controls, what to do with his hands and feet, how to regulate his engine, and other essential details. Since he cannot leave the ground his first concern is to steer the machine in a straight line, or as nearly straight as possible. This may sound simple, but remember that the steering is not done by means of the wheels on which it rolls, as you would guide an automobile. The wheels are not deflected and the steering, so far as right, left or straight ahead is concerned, must be accomplished by the rudder in the same way as if the machine were in the air. Of course the student is not concerned about up and down just yet, but he usually has troubles enough of his own as it is.

To the novice these *penguins* sometimes seem to be possessed of the devil. They take a sudden swerve to the right or left, apparently without warning or reason; then they begin to travel in circles and cut all kinds of wild and crazy capers. They seem to do everything but travel in a straight line or go where you are trying to steer them. A balky and fractious colt in the hands of an inexperienced driver has nothing on a *penguin* with a green hand trying to manage it. Perhaps you have had the same experience learning to ride a bicycle, wobbling all over the street and deliberately bumping into the very things you are trying to avoid.

Four *penguins*, driven by new students, turned loose in a ten acre field, will furnish more amusement to the spectators than a circus. It is a sight always enjoyed by everybody, that is—everybody except the students who are trying to manage them. But with experience and practice the student learns to anticipate their swerves and turns, he gradually accomplishes the knack of keeping them under control, finally masters the trick of driving them in a perfectly straight line, turning or circling them according to his desires or instructions and making them obey his will. He also learns the first essential lesson, that planes are controlled by gentleness and not by strength, by a soft and easy touch and not by sudden pushes or jerks. Then he is ready for the next step in his training.

After the *penguin* has been tamed the student is mounted on the "*Trois Pate*," or three cylinder Bleriot, with a normal wing spread of about twenty feet. This machine will fly but, the point is, the student must not let it do so. He is told to drive it in a straight line on the ground, learn to master it and *not let it fly* with him. Many times it does fly, in spite of instructions and the student's frantic efforts to keep his machine down. He may unwittingly hold the plane "in the line of flight" and if that happens the machine must mount—"it's the nature of the beast." Then, in his sudden panic, he generally forgets all about what he has been told and especially the right thing to do in that emergency. When this happens the usual or common result is that the prospective pilot gets his first crash, maybe turning a somersault, when, with a sudden dive, he hits the ground again. This sight is also highly amusing, a veritable rodeo of the air, and can be enjoyed because it is usually not serious.

It is one of the diversions of the older students, to see "the plane flying the pilot," up and down, up and down, like a bucking broncho, the student getting "rattled" and losing his head simply because, in his excitement, he overlooks the very simple expedient of shutting off his motor and bringing "the darn thing" to a stop. The fun loses none of its spice for the spectators because to them it may be a case of "We've all been there before, many a time, many a time." But time and practice solve these problems for the apt student and then he passes on to the next step, which is with the six cylinder Bleriot.

Now the student for the first time really experiences the sensation of sustained flight, and must learn to control his machine in the air. At first he just skims the ground, say six or eight feet above it, between two flags at opposite ends of the field, always under the watchful eye of his instructor, Then he mounts to twenty feet, increasing his altitude gradually as he gains confidence and masters control. Higher and higher he goes until he attains 1,000 feet with as much ease and assurance as when he was rolling along the ground during his final lesson on the *penguin*. Height loses terror for him as he becomes familiar with his machine and begins to get on friendly terms with it. He learns to make landings from different altitudes with his motor running and also with the motor shut off. This work continues, under varying conditions of wind and weather, with the height and duration of flights gradually increased until he has had a minimum of thirty hours in the air—thirty hours of actual flight. Then he is ready to try for his *brevet militaire*.

The *brevet militaire* is received after he has performed, to the satisfaction of his instructors and superiors, three tests or tasks. The first task is the making of two *petits voyages*; the second consists of three triangular voyages; and the third is the altitude test. All of these must be accomplished under conditions and within time limits imposed by his instructors.

Before starting on his *petits voyages* the student pilot is given a map of the country, over which he has never flown. As a matter of fact he probably has never been out of planing distance from his own field. A certain point is marked on his map; he is ordered to start from his aerodrome and fly to that spot, make a landing, have his papers signed by some person designated to perform that duty, as an evidence of the fact that he has reached the correct point, and then return to the starting point within a given time. If the test is not accomplished according to conditions and instructions it does not count. When the two *petits voyages* have been made satisfactorily he is ready for the second task, or triangle voyages.

Each triangular voyage consists of a three lap flight, covering a total distance of approximately 150 miles, at an altitude of not less than 3,000 feet, with landings at two intermediate points previously specified and marked on the map, and a return to the point of starting. These voyages must be completed within forty-eight hours, weather conditions or other untoward circumstances notwithstanding; otherwise it does not count as a credit. At the terminus of each of the two intermediate laps, which are towns with landing fields, the student pilot must descend, stop, have his papers examined and signed by the mayor or some other official, as a record of the time and place, and also have his barograph read and certified. The final record is, of course, made at the starting point, which is also the finish.

In making these voyages the student is thrown upon his own resources as soon as he leaves his aerodrome; he must use his wit and ingenuity to circumvent any difficulties or obstacles that may arise to hinder him. He is compelled to find his own way through the air, without any signposts to direct him and with no guides but his map and compass and such landmarks on the ground below as he can see and identify when atmospheric conditions are such that he can see the ground from the minimum altitude of 3,000 feet which he must maintain. So it is up to him to show his mettle and make good. Reason for failure must be very clear and concrete—a condition or circumstance entirely beyond his control or his power to foresee—in order to obtain another trial; excuses or *alibis* will not avail him.

Very often a student will become confused and lose his way on his trips away from the home aerodrome; he will fly here, there, and everywhere, trying to get his bearings or to identify some landmark, until in desperation, perhaps, he will make a landing at some likely looking spot. Many queer and amusing things have befallen the novices on account of these forced landings. One typical instance was that of a young American student out on his first voyage, early in 1916. He was lost, miles away in a direction nearly opposite to his destination, and landed near a small village on the Somme. Even though aeroplanes have become a somewhat common sight in many parts of France during the last two or three years, the sudden landing of one outside of an aerodrome has the same fascination for the small boy and the populace in general that it has in America. So the usual crowd collected when Adams (but that is not his real name) suddenly swooped down and landed near the village. His French was of the American public school variety—and he had not yet had much opportunity to improve it—serviceable to a certain extent, but not calculated to meet the demands of general or technical conversation in the heart of France,

Adams tried, with his scraps of poor French, to find out where he was and to ask for direction to his destination. About this time the mayor of the village, a veteran of the war of 1870, appeared on the scene and Adams turned to him, bringing to bear all the French that he could rake and scrape from his memory. But the mayor also failed to comprehend what Adams was driving at, while Adams was equally at sea over the mayor's speech. Things seemed to be almost hopeless, none of the crowd could speak English, and then a bright idea struck Adams! He had a smattering of German too, perhaps he could do better in that language and probably someone there would understand German.

But alas for his hopes! As soon as he began talking in that hated tongue the whole crowd began to bristle with hostility and the next thing he knew he was placed under arrest as a German spy, hedged about with a menacing circle of pitchforks, the business ends pointed his way, in the hands of the honest but determined country folks. He was marched to the village where, in due course of time, with the aid of his papers and the sign language, interspersed with a little good old American slang, he succeeded in persuading the mayor to communicate with the Commandant of the school. The Commandant was able to clear up the situation to the satisfaction of the mayor and Adams was finally given his direction and sent on his way, with profuse apologies and many expressions of good will.

He returned to his aerodrome, humble and in chastened spirit, and with two very fixed determinations in his mind: to improve his French with all possible speed and never again to speak a word of German back of the front line trenches.

To pass the third task, the altitude test, the student must mount to a height of 6,000 feet within a given time and maintain his plane at or above that altitude for *more* than one hour; then return and make his landing. The record of the barograph, which is sealed against possible tampering, is the proof of this test.

If, during any stage of the instruction, as I have outlined it, the student crashes more than twice (by a "crash" I mean any accident that seriously damages his machine) he is "radiated," that is, dropped from the rolls of the school and assigned to a "double command" or dual control school, where he may continue his training, if he still seems to be of promising material. Here he learns to fly with the aid of an instructor who rides with him in a two-seater machine having dual controls. But as this implies a lack of skill, or of the characteristics which go to make a good *chasse* pilot, all of the men who are ambitious to drive a pursuit plane very naturally strive to avoid that contingency.

The method of training in the double command schools is much the same as is followed in the American aviation schools and is done on the Caudron and Farman type of planes, which are used in the artillery regulating service. The student is given a limited amount of training on the dual control machines, then has his trial at solo flying, practicing landings, until, if he makes good, he is ready for his brevet tests, which are the same as those already outlined and are standard for all branches of the aviation service.

After the student has passed his brevet tests he has earned the right to wear the star and wings on the collar of his blouse, as evidence of the fact that he has attained that degree of efficiency. Heretofore he has been an "*Elève-pilote*" (student pilot); now he is a "*Pilote*" and is promoted to the grade of corporal, if his grade is below that at the time of taking his tests. But he is not yet a finished pilot, not by a long way.

From the Bleriot school he is sent to the Nieuport school, where the much swifter and more powerful 80 hp. Nieuport planes are used, with a landing speed of from 40 to 60 miles an hour, much faster than the Bleriot. His training continues here in a more intensified form; he flies at higher altitudes, makes steeper and more rapid ascents and dives that are swifter and more precipitous. Upon the satisfactory completion of fifteen hours of actual flight in the Nieuport he is given another altitude test in which a half-hour of sustained flight at a minimum of 9,000 feet is required. Even during his Nieuport training he is liable to be radiated to planes of another class if it should be his misfortune to crash more than twice. But having passed in this school he is not yet a *chasse* pilot; no indeed, the important part of his training, the real tests of his ability and courage are yet to come. A great many fellows, having passed the Nieuport school, flatter themselves that they are "regular pilots" and, I may as well confess, I had some such idea in the back of my head when I had progressed that far. But I was soon to find out how little I really did know. As I look back now I sometimes smile at my ignorance and self-assurance, but I was not alone in that error.

If the candidate survives the elimination process through the Nieuport school he is sent to Pau, the School of Acrobatics and Combat. Here he begins to learn what aviation, according to present day standards, really means, for at Pau he starts all over again, to a certain extent. That is, he unlearns most of what he has already learned in the former schools.

In his previous schooling his training and experience taught him to make a landing in a certain way, to bank at specified degrees, to climb at definite angles, and many other fundamentals. But now he must learn to disregard these methods. He is taught literally to tumble around in the air, to abandon control of his machine and then regain it, to make all sorts of dips, dives, loops, wing-slips and tail spins—in short, all of the spectacular "stunts" that look so hazardous but are so essential to the work of a finished *chasse* pilot. These are the "acrobatics" and they are well named, for they mean just what the word implies and the pilot doing them is indeed an acrobat of the air.

To the layman these stunts may appear as mere tricks or fancy frills prompted by the pilot's desire to show off, but any or all of them may be part of the day's work to a *chasse* pilot manoeuvring to get the advantage of position when engaged in combat with an enemy. Dangerous they may be to a man not thoroughly trained and with reflex not fully developed, but not necessarily dangerous when performed at sufficient height by the skilled aviator, which every man must be who hopes to survive for long in the contest now being waged in the air over the trenches of Europe.

Of all his days of training here is where discipline and obedience to instruction are most essential to the student. Failure to heed well what is told him and to do *exactly as he is told to do* will surely result in his failure to pass the tests demanded, if indeed it does not result more seriously for him. Without this discipline, which sometimes amounts almost to blind obedience, he cannot learn to do the things that are required of him—his reason will not let him. His only chance for success, or for life, in this school, is to *obey*, to do just exactly what his instructor tells him to do and in the exact manner specified, without question, comment or variation. He *must* place his confidence and his life absolutely in the hands of his teacher and carry out his instructions to the very letter or he had better not attempt to progress further as an aviator.

When I first went to Pau five pilots were killed in four days, right before our eyes. Accidents had happened before, plenty of them, and a certain number were to be expected—that is part of the game, one of the hazards every man must be prepared to take. But this was unusual, away above the average, and not calculated to inspire the students with the necessary confidence in their instructors and in themselves.

French Morane Parasol

Our Commandant knew that there must be a reason for this unusual record and he was not long in figuring out the answer. Then he called us together and gave us a good sound lecture.

"You men have no discipline," he said, "you lack the ability to obey instructions without question or comment. Why is this? You are all soldiers; you should be thoroughly disciplined and yet you are not! Remember this, it is discipline and obedience alone that will save your lives if you are to continue here. It makes no difference what you *think*—forget all of that—there is a reason for everything that you are told to do. Just remember what you are told to do and *do it*, then you will come out all right. First of all, get this fact into your head and always remember it; height is your friend, the ground is your enemy. Keep your instructions in mind when you are in the air and there will be no occasion for accident."

"Perhaps you are ordered to do a "*renversement*," he continued. "You get up in the air and start to do it when, *voila*, you feel a sudden blast of wind from a new direction; something seems to have gone wrong, you think you are in a *vrille* [tail spin] which you have been taught to dread. You lose your head, you try to think and then, *bang!*—you crash before you know it. If you do not lose your life, anyway you lose your nerve."

"None of you have discipline enough," he concluded, "but hereafter you will have. From now on each of you will do the *vrille* first of all, you will know what it is and how to overcome it; then if you come out of it you will do the other things after that."

These may not be his exact words, of course, but that is the gist of his lecture and we all took it well to heart. We all *did do* the *vrille* and everyone came out of it safely, for, it should be added, not a single serious accident happened for fifteen days, which was a new record up to that time.

René Simone, the "Bleriot Madman," has charge of the acrobatic school. He is a wonderfully capable pilot, one who flew when flying was really dangerous and on machines that were still in the experimental stage.

Before the student is sent up to do his acrobatics each manoeuvre, each stunt, is explained to him in detail by this veteran. The pilot is required to go over each step, each movement of his controls in their proper sequence, to rehearse the whole thing in the presence of his instructor. Then he "takes the air," climbing to a height of four or five thousand feet or more, to make the trial.

Sometimes the plane will hover over the aerodrome, a mere speck away up there in the sky. It hovers, circles and hovers again, seemingly for an endless time as the pilot is summoning his courage, screwing it to the point of daring to take that swift dive into terrifying space. It is the first great test of nerve, deliberately to fall two or three thousand feet perhaps. Still he waits, maybe; the suspense is awful, agonizing alike to himself and to his friends below who are watching him from the ground. Maybe he will begin his stunt, then before the plane really gets out of control he pulls it up, brings it back to control and hovers over us again. He is fighting the battle with self, summoning his pride, his courage, his determination to do what may mean swift death. It is a battle indeed, but he sticks to it, waiting for that moment when something seems to say "Go!" and he takes the leap. Down on the ground we are still watching, feeling the same emotions that are whirling through his brain; for we, too, have been in the same predicament and have experienced the same conflict that is waging within him.

Suddenly there is a shout, "There he comes!" He falls, he twists, turns, spins, down, down, down and then—O God!—he comes out of it. The breathless tenseness of those few seconds is dreadful; for he is our friend and comrade, dear to us who have shared with him our bread, our joys, trials and sorrows through many months. We are on hand to greet and congratulate him when he lands. His face beams, radiant with triumph and joy, for he has withstood the supreme test; he has accomplished the one great stunt. With this accomplishment comes the realization of something he was unable to understand before, that stunts are really the easiest of all flying to do, once you have won your own self-confidence.

A special course in formation or group flying is also given at this school. Previous to that time one has always avoided another machine in the air, giving it as wide berth as possible. But group flying is very essential in actual warfare, for it is very seldom that a machine flies over the enemy lines alone. To hold a compact and prearranged formation in the air with several machines in the group, flying at speeds varying from 100 to 150 miles per hour, is not easy. It requires strict attention to business on the part of everyone. Not only must the speed of each and all be regulated and controlled so that they may keep together, but they must all fly at the same altitude. There are many little things that may throw one or more machines out of formation; air bumps, which may elevate or lower a machine hundreds of feet in a very few seconds, add to the difficulty.

When doing this work a chief or leader of the group is named. He is the pivot of the formation. Each of the other members is then given a position, behind the leader and to the right or left, and he is supposed to keep it unless he falls out for some good reason. A rendezvous, some point in the air over a known spot or mark previously understood, is agreed upon. The planes leave the ground one at a time and head for the rendezvous. As each man arrives he circles around until the last one is in sight; then all circle until each one can fall into his proper position and, when all are in formation, the leader gives the signal and they head off like a flock of birds, one in advance while the others "follow the leader." Twelve hours of actual flight suffices for this training at the present time when the call for pilots is so urgent.

During this group training the pilot must learn the means of communicating with his comrades. The French system of signals is composed principally of certain manoeuvres, each having a definite meaning. For example: the chief "balances,"—see-saws his plane back and forth—as a signal for the departure or to attract the attention of the members of his group to him for any reason. If one of the group wishes to leave the formation on account of motor trouble, or for any other good reason, he leaves his position, comes alongside the chief and executes a *renversement*, or "Immelman turn."

The Royal Flying Corps has a more elaborate system. Each pilot is equipped with a "Véry pistol," which shoots balls of light of different colours, like a Roman candle. In this system, for example, a red light fired by the chief may mean, "I am about to attack," or "Close in," after an attack is begun. The same signal fired by one of the group may mean that he has seen the enemy. A red light fired from a single plane at a distance from the one observing it may mean "Help!" A green light fired by one of a group may mean that he is leaving formation on account, of engine trouble or because he is wounded. An intensive course in aerial gunnery is given at a special school. This part of the training is of so great importance that today the qualifications for a pursuit or fighting pilot call for 30 *per cent*, proficiency in flying and 70 *per cent* in gunnery.

The single-seater plane is essentially offensive. It is a plane of surprise, pursuit and attack. Its single gun has but one angle or direction of fire and that is straight ahead. Its only means of protection against an enemy's attack is in its agility, its quickness of movement in manoeuvring for position and its speed in flight. It cannot sustain a persistent or long drawn out duel. It can only deliver its attack quickly by thrusting at the vitals of the enemy, securing and keeping the advantage of position as much as possible. If this attack fails, if he is outmanoeuvred, outnumbered, or if his ammunition is exhausted, the pilot has no alternative but to retire from the combat and, if possible, retreat to the safety of his aerodrome.

Therefore, while the success of attack depends primarily upon the skilful acrobatic manoeuvres of the attacking plane, shooting ability is obviously very essential to enable the pilot to take advantage of any favourable position which he may secure. As he may be able to hold that position for only a very few seconds he must grasp the opportunity quickly and make the most of it. As a matter of fact, great success comes only to the few exceptional and superior pilots. It is impossible to create a large body of aviation geniuses by training or any other means, just as it is impossible to make great artists out of all who study music. A certain amount of proficiency can be attained by training, practice and experience, but the phenomenal pilot is the one whose peculiar talents and temperament cause him to be particularly adapted to the work and who, for that reason, takes to it quickly, naturally and easily.

As a consequence, in training the whole student body or general average of aviators, greater importance must be attached to gunnery than to acrobatics. There is strength in numbers and with enough gun, popping at the enemy a certain percentage of hits are bound to result. Of course, the average enemy pilot is no better, if, indeed, he is as good as the average man on the side of the Allies. Skill in marksmanship is more easily attained through training and practice by the general run of men from which pilots must be selected than great ability in acrobatics, though, as has been said, one is really the complement of the other so far as successful individual attack is concerned. Recognizing all these facts, a very careful and elaborate system of training in gunnery has been instituted in France.

In the French system the theory and practice of gunnery are taught simultaneously. The student receives daily instruction in the nomenclature of machine guns and the parts thereof; in assembling and disassembling the various types in use. He engages in small arm practice, both with pistol and rifle, with stationary and moving targets. This includes trap shooting also. He is then given machine gun practice at various ranges, firing first from a stationary base at moving, disappearing and surprise targets. Then he shoots from a moving base at the same targets and so on, with numerous variations and changes, until he is qualified to shoot from the air.

He begins his gunnery practice in the air in a two-seater plane, shooting at stationary marks, at objects floating on the water, at balloons floating in the air, and at targets towed by another plane. All this shooting is done with a movable machine gun, one so mounted that it can be turned from side to side and up and down. Good marksmanship is rather difficult to attain because of three varying factors, namely: the speed at which the gunner is travelling, the speed at which the target or enemy is moving, and the speed of the projectile. However, sights for the guns have been perfected which automatically make allowance for these variables and with knowledge of their purpose and practice in manipulating them, good marksmanship can be developed.

After attaining reasonable proficiency in this practice the pilot begins his gunnery training in the single-seater machines. In these planes the machine gun is mounted in a fixed or stationary position on the motor cowl; it is pointed up, down or to either side by guiding the plane at a corresponding angle. In other words, the gun shoots as the plane flies, straight ahead only, on the axis of the plane or along the line of flight; it cannot be moved relative to the plane. In a fighting plane the gun fires through the propeller, as will be explained later on. The same sort of targets are used for gunnery practice in the single-seater planes as were used for the two-seaters. Proficiency in this work is quite difficult to acquire; chasing a balloon two feet in diameter and hitting it with a 30-calibre bullet is a rather difficult feat.

The course of gunnery training consumes two weeks of time and when completed the Scout Pilot is at last ready for the front. Other branches of the service and their courses of training will be touched upon later.

During the courses of training already outlined a special course of map reading is given to the pilots. This consists in locating and checking up rivers, roads, towns, forests, hills, etc. An elementary course in astronomy is also given; the use of the compass is explained and its points must be learned. The meaning of the magnetic pole and its influence on the compass is elucidated. The principal stars and constellations are identified and learned so that the pilot may take his bearings from them if he should become lost at night. The position of the sun in the heavens at various hours of the day and seasons of the year is explained.

The winds and their peculiarities are also memorized. Winds from different directions have different characteristics. For example, winds from the west, in France, usually move in the higher air *strata*, or they are strongest in the higher altitudes. Winds from the east have characteristics the reverse of west winds; their circular trend is toward the south and south-west. North winds, on the surf ace, trend toward the west up to about 5,000 feet and are strong; above that altitude they are milder. Generally speaking the higher one flies the stronger are the winds which he encounters. As a rule winds have a tendency to blow clockwise; that is, they have a circular sweep, turning as the hands of a watch turn when it is lying face upward.

There are four distinct kinds of maps used, differing according to scale. As usual, the top is north, the right hand east, etc. The types and scales used are, 1 to 5,000; 1 to 10,000; 1 to 20,000; and 1 to 40,000.

1/5 M. Special trench map, showing all enemy positions according to the latest available data; gun emplacements, the different lines of trenches with their communicating trenches, machine gun stations, etc. These are made from photographs which are being taken constantly from aeroplanes, new maps being issued every week or two. They are also used by infantry commanders in making attacks, so that a sufficient number of men may be assigned to the duty of capturing each enemy gun which may be in the line and range of attack.

1/10 M. Trench map, showing both Allied and enemy lines, in less detail than the first named. Enemy positions are marked in red, Allied positions in blue. These are used by division commanders in planning and executing attacks.

1/20 M. Artillery regulating map, showing cantonments, reserve depots, ammunition dumps, railroad stations, hospitals, gun batteries, etc. These are used by planes in regulating artillery fire and for bombing any of these points in the enemy's territory.

1/40 M. Reconnaissance map, showing an entire sector, with streams, roads, villages, towns, railroads, etc. These are used by scout pilots and photographic planes in making aerial photographs from which the larger scale maps are drawn. Each pilot carries one of these maps fixed on his plane in front of him so that he may locate his position and avoid getting lost.

The course in this work which is given to pilots is called "Map Reading," but the much more thorough course given to observers is known as "Orientation."

The observers and photographers have their elementary training on the ground. They, too, learn map reading so as to take pictures over the lines that will fit into the great plan at headquarters. For the most part they are officers. There are a few non-commissioned officers in the lot. They are also trained in telegraphy, radio work, etc. Frederick Zinn is the only American in this branch of service in the French army.

When an observer has completed his ground work he gets about a month's training in a two-seater plane. He then learns how the world really looks when he's a mile or so above it. He learns to gauge distance. He studies light effects and most of all he studies shadows—the most effective *camouflage* in the world.

Camouflage has now developed into a science and there are many things to fool the new hand at the game, some of them of natural origin and some of them of Boche origin. The art of camouflage is being adapted to the camera as well as to the human eye. It isn't so long ago that canvas screens bearing painted landscapes were quite effective and deceived the keen eye of many a pilot. Now the best all around device consists of long nets, similar to a seine or a tennis net. These are woven of green cord with strips of green and black cloth strung to the meshes. The strips are about one inch wide by ten or twelve long. They flutter in the slightest breeze and give exactly the effect of a field of grass, underbrush or trees, according to the distance the observer is away from them.

Aeroplanes and cameras, especially cameras, have taught both sides the value of natural shadows. Nowadays you will seldom find an ammunition dump, a supply depot or a barracks that is not situated at the border of a wood or near a line of trees where but once a day it will be unprotected by the shadows. That is at noon, when the sun is directly overhead, casting only a downward shadow.

Long ago the aeroplane stepped right up behind the artillery as a means of preparation for the attack. Concentration of planes no less than concentration of fire is a prelude to an assault. Not only must the advance work of mapping be done, but the enemy's planes must be absolutely overwhelmed by force of numbers. Not a hostile scout must be able to penetrate above our lines to spread the news of what is going on.

Of all the signs by which the Tommy and the *poilu* know that an assault is in preparation, the arrival of great numbers of aeroplanes is the most reliable. The crack *groupes de combat* are withdrawn from other sectors and sent to that in which the new hostilities are to begin.

Two or three weeks before the date set for the actual assault the monster flocks come soaring through the sky. As many as five complete *groupes de combat* may be sent to one sector in preparation for an assault. At daybreak the scouts go whirring up in pairs and streak off, each pair to some appointed German town.

They check up all the trains in motion behind the lines, the length of each, the direction in which it is going, etc.; all troop movements, all fires and all other military information that comes to their notice. Within half an hour after daybreak the staff knows precisely what the situation is behind the German lines and is able to guess pretty shrewdly what the Boche plans are for the day. The actions of our own troops are regulated accordingly.

It was just such reconnoissance as this which made possible the redemption of the forts at Verdun in the summer of 1917.

The beginning of the last Battle of Verdun was really in Flanders. It was in July and the early part of August. The entire country had been mapped, the artillery was keeping up an incessant fire, day and night. "Hurricane fire" gave birth to its name here. On the eighth day came the attack. It was a complete success.

Then it began to rain. Just a little murkiness that hid the stars at first, then a drizzle, and then a downpour. It rained for three days. The planes were useless; the army was blind and couldn't move. For three days the attack was held up. We could not move our large guns through the mud; we knew not where we were outnumbered—we could only wait.

On the fourth day the dawn was clear. We started out on reconnoissance, thundering up in pairs at the first streak of light like so many huge partridges. I was fortunate enough to be one of a pair whose destination was a certain town, Thourout, that was a strategic railway centre.

While we hovered over that place for an hour we counted fourteen long trains that pulled into the station. Each train disgorged Germans. Troops were being concentrated there at a tremendous rate. We saw movements of convoys on the roads in daytime, a thing before unheard of, for as a rule all traffic moves only under cover of darkness in the war zone.

By the time we had counted fourteen trains the sun was beginning to climb, and we started straight back to our lines. We reached them without incident, except for the usual salute of shrapnel, and reported what we had discovered.

Although the day was fair, the attack was not prosecuted. It was evident that the Boche was ready for us. The advance and the supports were left to dig themselves in and hold what they had won, but the whole mobile reserve was packed into trains and started at top speed for Verdun.

Everybody now knows what happened then, but not many know what made it possible and why the attack was switched from Flanders to Verdun. It was the blinding of the eyes of our army for those three days that prevented our following up our victory. It was the fact that we recovered our sight before the Boche did that let us win at Verdun.

CHAPTER 4

Air Fighting

The Allied flying men hold the supremacy of the air on the Western front today by sheer will power. It is the combined power of a thousand individual wills, of their daring, their skill and their devotion. But most of all it is their determination and confidence that keep the Boche back of his own lines in the daylight hours, when he can see the vital things. They are also the ones who permit the Allied generals to know, through their flying eyes, from hour to hour, of all that goes on behind the German front.

Armament today is equal. The Boche planes have machine guns as good and trustworthy as ours—and no better. They carry the same number of rounds of ammunition. Their planes are as fast. No new wrinkle, no new trick, no new device can be kept secret by either side for more than a few days after it appears at the front. This is due not only to the number of our planes brought down back of their lines, but to the German spy system—spies check up our planes as well as our troop movements. So, in equipment, we are about equal. It is probable that for the last few months we have had more planes than they, but they are always able to concentrate equal or greater numbers against us when they make the effort.

Air supremacy is not to be determined by counting noses or weighing guns. It is the sum of a great number of details—and each detail is a combat between an Allied and a Hun airman, hundreds in each twenty-four hours. We control the air because we win most of these combats—because the Boche fears us, fears to come over our lines, fears to take the risks without which he cannot make his work of great value to his General Staff. He has not given up the fight—far from it—he fights desperately to defend his own ground and to invade ours, yet day by day we beat him back, and day by day we go where it is most needful.

Victory in the combats on which all this depends comes down to a question of the relative flying and fighting ability of the pilots. Given equal courage—and the Boche aviators have that, whatever may be said of the men on the ground when they are thrown on their individual initiative—it is the flying that counts most. For each combat is a contest in air manoeuvres in which the winner lives.

To approach within a short distance of the enemy and to fire with certitude is a delicate thing, and one that requires much experience. There must be no swerving away from the enemy—you must often nearly run him down—and it frequently happens that when a collision 20,000 feet above ground seems inevitable it is the one whose nerve lasts a few seconds longer that reports back to his hangar. It is this spirit that carries the Allied fliers over those last few inches, together with the nerve and dash that disconcert and overawe the enemy, that give them the supremacy of the air.

Nor does this mean contempt of the enemy—that is fatal, and often has been. Fritz in the air is always dangerous, and one comes to regard him with a very real respect—not for his cause, but for his ability. There are too many Allied planes that do not come home, too many pilots marked "missing," for any man who has been long at the front to have an idea that he can give Fritz anything less than the best he has. But our best pilots know that when they do hang on for the last inch they will get it and will win.

One of the most thrilling and spectacular air duels ever witnessed started because one of our greatest aces, Navarre, had underestimated the Boche opposed to him. It was in the region of Dixmude, and Navarre swooped at him a little too carelessly. The Boche gave him the hardest battle of his career. Up and down, in loop after loop, in spiral, dive and twist they flew, slipping, turning and dodging, each striving for the fatal second when he could press the trigger on which his finger rested. Time and again there came a burst of fire, but always it was too late and the man at whom it was directed had just twisted out of line. Finally the Boche tried to drop out of the contest by planing for his own lines, but Navarre followed him, cut off his retreat and forced him to land behind our lines. Out of respect for his adversary Navarre landed beside him and found his opponent was a boy of seventeen. The German had quit the duel only because his gasoline reservoir had been punctured and his ammunition exhausted. The battle was one of the longest on record, lasting nearly five minutes.

It is on the *chasse escadrilles* (hunting squadrons) that the supremacy of the air rests, and with it the security of all that is going on, in and behind the Allied lines. They are the scouts and fighters; their chief mission is to keep the enemy planes from invading our territory, and this they do by an unceasing patrol of the lines, giving battle to any Boche who presumes to venture from his own domain. The patrol is no hit-and-miss operation; it is as scientific, as steady and as careful as the picket duty in the front trenches, or the British destroyer procession across the Channel. It is they who have kept the Boche from repeating his cowardly raids on Paris, where he set out to duplicate his murders of women and children in London.

They also protect the planes engaged in regulating our artillery fire, and they have other interesting duties which I will discuss later on. Their main activity is to fight—they are the duellists of the air, and it is this work that I will describe now.

The details of training have already been told. It is sufficient to recall that for this work only the very best pilots are assigned, as the delicacy of the duties call for many and rare qualifications. The pilot must, of course, be imbued with the highest ideals of duty and self-sacrifice. He must have undaunted courage and perfect confidence; he must be cool under any conditions that may arise. To attack successfully he must always do so with the conviction that he will be the victor.

To help in recognizing enemy planes each *escadrille* is supplied with silhouette photographs of all known types of Boche machines, in all possible positions. With study the pilot soon becomes able to recognize these at great distances. He has to depend far more on the types of machines than on the markings which are supposed to show the nationality of the plane, because the Boche is using a great many machines which are carrying the *cocarde* of the Allies. Usually these have a very small black cross painted in the centre of the *cocarde*, but this cannot be seen until very close, though under powerful telescopes it is plain enough to keep the Boche gunners from firing on their own machines. A pilot who approaches unwarily, in the belief that the machine is friendly may pay dearly for his fault.

In the Flanders district the recognition of planes is particularly difficult, because of the great number of types in use there. Beside, the usual familiar types of Boche machines, and the French ones, there are planes of the British Royal Naval Air Service, the British Royal Flying Corps (different organizations with different types of machines) , the Belgian Aviation Corps, and the Portuguese army planes. Their types and sizes vary so much it is easy to make a mistake, and this happens quite frequently. Once a member of a famous French *escadrille* fought a duel at 15,000 feet with a member of the Royal Flying Corps. They did not discover their mistake till the battle had become a draw, and they had parted without injury to either. Later both were reprimanded by their officers, and there were mutual apologies. The use of the Allied *cocarde* by the Huns increases the danger of such mistakes, as well as making it harder for an Allied pilot to know whether a machine which he may sight is friend or foe.

When an enemy plane has been sighted, manoeuvring for the attack begins at once. One should take advantage of every bit of natural cover, in an effort to catch the Hun unawares. If the day is clear, for instance, the pilot should try to get between his enemy and the sun. The sunlight will blind the Boche and the attacker can get within a few feet of him before being discovered. It is almost impossible to see a machine diving from above with the sun behind it. The use of clouds for both attack and escape is of course very frequent. This can be done best when the clouds are large fleecy white ones, scattered about the sky. One may fly above them, catching occasional glimpses of the earth to avoid being lost, and one has a gallery seat for the whole show, seeing all planes in the air beneath.

As a protection against attacks from above most planes of both sides are camouflaged on the upper side of the wings. The difficulty of creating a successful disguise for a plane that will work under all conditions is great, since the background changes so frequently. Still, the camouflage may be designed for the kind of background that is most frequent and it helps just that much. Attempts are also made to colour the under sides of the wings so as to make them blend with the sky—pale blue, pea green and aluminum usually being employed for this purpose. This is to furnish a poorer target for anti-aircraft guns and also to enable one to drop down on an enemy from above without being seen.

Aerial combat involves a series of *actions* by means of which the combatants strive to make the best possible use of the equipment and arms at their disposal for the destruction or rout of an opponent.

The elements involved in actions fall under two general headings, namely:

(1) *the tactical elements of combat* are those whose end is to take the best advantage of the most favourable time, position and

formation;

(2) *the technical elements of combat* are those depending upon the qualities and equipment of the plane.

To achieve victory both the tactical and technical elements must be combined and employed.

In an aerial duel or combat, in which it is assumed that the one attacking is a scout or pursuit pilot, the first things the fighter strives for are the greatest possible advantages for himself. These advantages are: to surprise his adversary, to be flying at a higher altitude at the time of attack, and, when the fighting is done by groups, to be in the best formation.

There are, and can be, no definite rules laid down to govern any or all attacks for the simple reason that all the surrounding conditions may not be the same for any two. The particular circumstances of the encounter, the relative qualities of his plane and those of his adversary's machine, which the pilot must weigh and estimate, are the deciding factors which govern the plan of action. This is where consideration of both the tactical and technical elements enter. The tactical cannot be executed with hope of success unless the pilot possesses the proper technical elements in his plane.

There are four of these technical elements, namely: *speed, manageability, armament* and *altitude capacity*.

Some planes demand technical elements totally different from others, depending on the principal work they have to perform. As they are designed for a particular purpose so they must possess some qualities not found in others of a different class and lack some elements which are a part of the others. It is impossible to build a plane that will possess all of the good qualities of the several classes just as it is impossible to construct a naval vessel which combines all of the advantages and characteristics of a destroyer, a cruiser and a super-dreadnaught.

The Spad and the Nieuport machines constructed by the French undoubtedly combine the four technical elements to a higher degree than any others which have been designed up to the present time. But as changes take place so rapidly, because of the great amount of experience gained and the highly intensive study which is being given to all of the elements which contribute to greater efficiency in aeroplanes, these types may be superseded before long.

The artillery regulating, the photographic, and the bombing planes possess many qualities of technical superiority, as compared with the

Spads and Nieuports, especially as to armament. The peculiar work for which each of these types is designed demands large carrying capacity, wide radius of action, and comparatively easy and safe landing qualities. These elements are not to be found in either the Spad or the Nieupofct machines.

Speed is very important, in fact, absolutely essential. For it is speed that enables a pilot to overtake his prospective adversary and thus gain the first advantage of surprise in the impending attack. The actual speed of the Spads and Nieuports may vary but they will travel, say, up to 150 miles per hour. By actual speed, I mean the maximum speed at the maximum altitude. As all automobile drivers know, carburetion varies to some extent under different atmospheric conditions and motor efficiency is affected, to a greater or less degree, accordingly. This is particularly true with aeroplane motors when flying, as the engine may encounter many changes in atmospheric conditions during the course of a flight. In the rarefied atmosphere of the higher altitudes the mixture is quite different from that which is produced by the same carburettor on the ground and as a consequence motor power and efficiency is materially reduced. But the speed of 150 miles per hour is near enough to be approximately correct at an altitude of 17,000 feet; at the lower altitudes and under favourable conditions the speed is much greater. This speed could be increased still more but it would be at the sacrifice of some of the manageability of the machines and also of some of their "safety of landing" factor.

Manageability, in the pursuit planes, is almost unlimited. They have been designed with that particular element in view and to stand up under the most racking acrobatic work. This has been possible because of the large amount of exact scientific knowledge of stress, strain and pressure, as applied to aeroplane construction, which has been gained. The quality of manageability depends upon the well-balanced and proper distribution of the stresses and strains to which the plane may be subjected, a sufficient factor of safety being allowed without excess weight. If an accident occurs through the breaking of some part of the plane it is usually due to defective material and not to fault in the design. The best insurance against such accidents is constant and careful inspection. There are three forces to be considered in designing a plane; the force tending to make it climb, the force tending to pull it down, and the force resulting from the use of its controls.

Armament for defence or attack is necessary in all planes which are

FRENCH MORANE PARASOL (FRONT VIEW)

THE FASTEST MACHINE MADE, WHICH CLIMBS 16,404 FWWT IN FIFTEEN MINUTES

used for war purposes. Early in the war a rifle or a shot gun was the only weapon carried. Pilots, both Allied and Boche, saw each other in the air, passed each other while going back and forth, and each took an occasional shot at the other when he got a chance. But that was about all. Slowly, progressively and independently each began experimenting with machine guns.

Early French planes, the Farman and Voisin "Pushers," had their propellers in the rear, behind the pilot. When machine guns were placed in these planes they were mounted in the front of the car. The Boche planes, for the most part, were tractors, with the propellers in front, the machine gun and pilot being in the rear. Therefore the blind angle of fire, as between the two types, was directly opposite and the Frenchman, in order to fire at the German, was compelled to "get behind him," or *vice versa*. Many desperate encounters at close quarters ensued but little material damage was done on either side. This was partly due to their ignorance, at that time, of the three speed factors, already touched upon, and because of the vibrations caused by the light guns.

When the advantage of speed was fully recognized and the superiority of the tractor type was established experiments were made at mounting stationary guns of the Lewis type above the upper plane, so as to fire over the propeller. This method avoided the vibrations, due to the firing of the machine gun, because they were absorbed by the entire plane. But as the magazines contained only 47 cartridges, which were fired at the rate of 300 or 400 per minute, and as the guns were inaccessible for reloading, the supply of ammunition was too small and too quickly exhausted to make this arrangement of much value.

Following this expedient, Garros, one of the most notable of the French pilots, began a series of experiments of firing *through* the propeller. The machine gun was mounted on the cowl of the motor and was fired at its normal rate. In order to avoid the possibility of the shots shattering the propeller blades, Garros covered the blades with thin sheets of hardened steel to deflect any bullets that might strike them. On account of the high velocity of the bullets and the rapid speed of the propeller about 8 *per cent*, of the bullets struck the blades. Those that passed the blades were good but those striking the plates were wasted. Another objection developed and that was that the bullets striking the blades had a retarding effect upon the propeller with a consequent loss of flying speed amounting to about fifteen miles an hour. However, enough was learned through these experiments to demonstrate that the scheme was good and practical.

Then it was decided to synchronize the gunfire with the revolutions of the propeller by connecting the loading and firing mechanism of the machine gun with the cam shaft of the motor in such way that the firing of each shot would always occur at an instant when neither of the propeller blades is or can be in front of the muzzle of the gun. With a propeller having two blades and revolving at the rate of 1400 revolutions per minute, the blades pass the muzzle of the gun 2800 times during each minute. Therefore the interval between blades is only about one forty-sixth part of a second. So the synchronizing job is rather an exact one. However, it was accomplished and today planes with propellers revolving at the rate of 2200 revolutions per minute, giving intervals of only one seventy-third part of a second, have machine guns that shoot between the blades.

At the present time some guns mounted on a plane carry a belt containing from 500 to 1,000 rounds of ammunition, in comparison with the original 47 shots in the magazine of the first machine guns used on planes.

The armament on a *chasse* plane is essentially offensive for the reason that, as already stated, the gun is mounted in a fixed position and the direction of fire can be changed only by altering the plane's line of flight; in other words, the gun is pointed by pointing the plane at the object which the pilot is endeavouring to hit with his shots. Therefore it is manifest that in order to make the gun fire effective the plane's line of flight must be in the direction of the enemy; you must fly *at him* in order to hit him. As this is done only when you are making an offensive thrust the armament on the scout plane cannot be used for defensive purposes.

The two-seater planes, Caudrons, Sopwiths and Farmans, used for artillery regulating, aerial photography, and bombing, have both offensive and defensive armament and therefore, in this respect, have a technical advantage over the pursuit planes. They have a forward machine gun which fires through the propeller, to be used when, in self-defence, they must attack, and a rear gun, which is mounted on a moveable base, to repel attack from that quarter. But, naturally, they have not the speed or the manageability of the scout planes. Therefore they are not intended for offensive work and are equipped with armament primarily for defensive purposes only.

The three-seater planes have an armament which is essentially defensive. The French "Letours" and the German "Gothas" are of this type. They have two motors and propellers mounted in front and hence are tractors. They carry three guns, all of which are mounted so that they are moveable. One gun is in the front part of the car for shooting forward; the other two are in the rear, one shooting over the top of the fuselage and the other shooting below through an opening or tunnel in the bottom. Therefore they have a very wide range of fire and are exceedingly hard to attack successfully. Nevertheless, Guynemer brought down several of them during his career. A scout pilot in attacking them must bring all of his acrobatic ability into play. The right tactics would be to wear out the rear gunner, who operates both the upper and the lower gun, forcing him back and forth from his upper gun to the one in the tunnel, manoeuvring so as to harass him and keep him jumping from one to the other until the death shot is fired.

These three-seaters have an immense wing-spread, necessary to give them their great weight carrying capacity, and they have a long cruising radius. It is the German Gothas that have bombed London so frequently and the French Letours which have retaliated on German cities, but the German censorship has kept the results of the French raids from the press.

There are four types of machine gun ammunition used at the present time in aerial warfare: the ordinary, the incendiary or tracing, the perforating and the explosive. The ordinary cartridge is the same as that used in the same type of gun in the trenches. The incendiary or tracer bullet is steel jacketed and has a hollow core which contains an inflammable material of a phosphoric base. This material ignites when the cartridge is exploded and burns while the bullet is in flight. They can be seen by the gunner as they speed through the air like a small ball of fire. They are used to ignite any inflammable material which they may strike, as, for instance, the gasoline tank of an enemy plane; also to correct the gunner's aim, as, by watching or "feeling" with them, one can see how close to the mark he is shooting.

The perforating bullet is made of a special, hardened steel which is enveloped in a coating or jacket of bronze. They will perforate metallic parts, motors, etc., the bronze acting as a lubricant when they impinge upon metal, the principle being the same as the armour-piercing shells used in naval guns.

The explosive bullet is really a small shell in shape, form and composition. They contain an explosive charge which is detonated by a small piece of metal which delivers a blow upon a fulminating capsule the moment the nose of the bullet impinges upon any resisting surface. They are used especially against dirigibles, sausage balloons or other gas bag air craft. Their use is not common except on special missions involving a purpose which requires a missile of this sort.

The ammunition belts used with the machine guns on the present-day pursuit planes usually carry three kinds of cartridges, distributed in the following proportions: one ordinary cartridge for each tracer or incendiary, and one perforating for each two of the other kinds. (Ordinary, tracer, ordinary, tracer, perforating; ordinary, tracer, ordinary, tracer, perforating, etc.)

Altitude capacity, or ability to maintain great height, is the fourth and last technical element involved in aerial combat. Superior altitude is equivalent to a reserve of speed in a pursuit plane. This is especially true when a pilot is diving down upon an adversary, as is apparent. Moreover, he who has superior altitude and equal or greater speed may force or decline combat at will. If he is above the enemy he may dive down upon his victim very quickly and easily while it may take consider- able time, distance and manoeuvring for the enemy to climb to his level—if, indeed, it is at all possible or practical.

Taking all of the variable elements touched upon into consideration it is apparent that a plane may have one technical advantage, armament, without being able to secure a tactical advantage. For example, a Voisin bombing plane is equipped with a 37 millimetre rapid fire gun which was, at one time, superior to anything the enemy had mounted on their planes. Yet this plane, having neither speed nor altitude capacity, could be attacked with ease and comparative safety by a swift pursuit plane. On account of its slowness it could not pursue, and on the other hand the scout plane could, if for any reason it elected to do so, climb a thousand feet higher and refuse combat. The Voisin could not follow because it lacks ability to attain the higher altitudes.

Thus it is a combination of the tactical and technical elements that should decide the issue of an aerial combat. Yet it is the pilot's appraisal of the relative value of all of these elements possessed by himself or his enemy, his ability to decide whether or not the margin is in his favour, and his quickness in making use of any advantage or opportunity, that really counts.

Every type of plane, whether friend or enemy, has its peculiarities; its dead angle of fire, its blind spots, its maximum and minimum fire arcs, its limit of speed, etc. Every pilot also, both Allied and Boche, has his individual characteristics, peculiarities and methods so that his identity may become apparent to the experienced eye. So it is study, practice, observation, experiment, and experience that can be gained only in actual combat, which enable a pilot to decide just what course of action he should follow in the exigencies of battle in order to insure his safety and at the same time, if possible, achieve success or victory.

The modes of attack, as has been stated, differ according to the type of enemy machines. If by good luck you catch a Boche isolated and the plane is a single seater, the favourite mode of attack is to dive down from behind and *chandelle* upward, slightly under and behind him—"getting under his tail," we call it. Here you can pump lead into him at will, as he carries but one gun—mounted for- ward. He must turn his plane to fire back at you. Another mode of attack, used when you meet instead of overtaking the enemy, is to dive head on, then execute a *renversement*, bringing you up and behind the enemy, pro- tected by his stabilizers, where you again have the advantage. The best position for this form of attack is known in French as "*de trois quatre en avant*" (from three-quarters in advance). Diving from that position, the Hun pilot cannot fire at you without shooting through his own wings—if he is a two-seater—and the *renversement* places you exactly under his tail.

The combat, of course, does not often work out in this simple way. The enemy pilot dodges to prevent your getting under his tail, and the duel then becomes a series of acrobatic manoeuvres, each man trying to get the position of advantage. Each man tries to get on the nerve of his adversary by keeping him under fire as constantly as possible, and at the same time protect himself by *virages*, slips and loops, so as to keep out of the other man's line of fire. The combat does not last long—a few seconds—when one man usually loses his nerve, becomes demor- alized and tries to escape.

If one is surprised in the air by an enemy he should never dive but always climb. If the attacker follows you should spiral upward, keeping him always inside the circle and in sight, and watch for an opportunity to dive at and under him. If, however, you see the enemy diving at you, it is good piloting to keep directly below him, as nearly as possible, and go through the same manoeuvres that he does, only at a lower altitude. If he dives, you dive; if he *virages*, you follow; if he climbs, you go up after him. If you do this he cannot bring a gun to bear on you, and there is always a chance that he will make a slip and let you get under his tail. If this does not happen he will soon give up and leave you.

The two-seater (*biplace*) machine is a more dangerous adversary, in that it is better protected by gunfire, though it has not the agility of the smaller machines. The two-seater is strongly guarded at the rear, having a machine gun there mounted on a swivel, fired by the observer, as well as the usual fixed gun in front, fired by the pilot This greatly reduces the "dead areas of fire" and makes it possible to come at the machine only from directly above, or behind and below. The new German Gothas have further decreased this dead area of fire by mounting a third gun on the floor of the *nacelle*, which is fired through a tunnel in the fuselage, so that the position "under the tail" becomes one of deadly danger, instead of an advantage.

To drive an aeroplane is nothing at all—all kinds of men and some women make good aviators—but to pilot it as one must in an aerial combat, is extremely difficult. A pilot must be able to do with his machine all that he wishes. He must make it respond to his every emotion, it must be a part of him, as the horse becomes part of an experienced rider. In fact, to enter the Royal Flying Corps of Great Britain a knowledge of horsemanship is required. He must be able to do the most complicated acrobatic feats automatically while his mind is on other matters. He must never be occupied for a second with the handling of his machine. That must always be a secondary consideration—a reflex, subconscious. This is only acquired by months of constant and methodical training, and to it too much importance cannot be attached.

This is in addition to his knowledge of methods of fighting. This, too, must be perfect, for it is a thing that changes almost from day to day. As our own or the enemy's pilots develop new tricks and modes of attack or escape, principles that are right today may be wrong to-morrow. It is possible the Boche has learned an answer to them. Also the pilot must consider all the different types of machines with which he may come into conflict There are many of these—every once in a while the enemy will produce a new one and each machine must be met and attacked in a different manner. There are different types used along different sectors of the front, and a pilot transferred to a new territory will have trouble for days before he gets acquainted with the new conditions.

Thus the science of air combat is almost an exact thing. Few movements have not been tried; few have been found for which no answer is possible. Each manoeuvre, to the skilled pilot, calls for a definite response; each position in relation to his enemy indicates what his

French Morane Parasol (side view)

This machine is equipped with a 170 horse-power motor

next move—and the Boche's—should be. In fact, the science has become in many ways like that of fencing, where each of the opponents counts. on the other doing the expected thing and attempts to win by catching the other off guard, or fighting with a little more dash and power than his enemy.

It is possible that his confidence in this code was the cause of Guynemer's death. He particularly enjoyed outflying and outmanoeuvring his enemies, and then, when their nerve was shaken, administering the death stroke. The Hun pilot who brought him down was a novice, as is shown by the fact that he was himself killed within a few days, and by his boasting letter to his mother that "now she need have no fear for him, since he had conquered the greatest of all airmen." But novices do the unexpected thing, and it is possible that this youngster, instead of replying to one of Guynemer's manoeuvres, by "executing a position of advantage," as the French ace expected, became excited and did "the wrong thing," according to the book, but one which caught Guynemer off his guard.

Let us now start with a French patrol on the day's work. These *chasse* patrols always work in groups of six or more, except the aces, who are sometimes allowed to fly in pairs. To fly alone is suicide, and there must be several planes in close touch to assure any degree of safety. Even Guynemer never thought of flying alone during the later months of his life; though it was different in the early days. Guynemer's *escadrille*, N-8, was composed of some of the best pilots of France and teamwork was their especial strength.

The fighting planes leave their aerodromes at one or two minute intervals. Each pilot as he leaves the ground reaches out and "arms" his machine gun by swinging a lever that projects from the right of his windshield and pushing the first cartridge in his belt under the hammer. The guns are never loaded when on the ground, as a jar might set them going, so the first duty of a pilot in going up is to arm his gun just as his last duty before alighting is to disarm it. Each pilot has a single belt of cartridges (200 rounds) coiled in a little box at the right and under his gun. That is his supply for the trip; if he exhausts it he must return to the hangar, if he can.

After leaving the hangars the planes proceed to some agreed rendezvous in the air, above some landmark. There they whip into formation. The chief of the patrol, after the group is formed, slowly balances—that is, see-saws his machine as a signal for departure—and they are off on their patrol of the lines. Each pilot in the group must watch and follow every movement of his chief. Recently, planes have been equipped with a radio receiving outfit, and are notified while in flight of any concentration of enemy planes in nearby sectors. Upon receipt of such news they start immediately for the scene and give battle.

Sometimes it happens that some man in the second or third position in the group will be the first to see an enemy plane. In that event he immediately leaves his position, flies up beside the chief and balances his machine. He then, for the time, becomes the chief of the group. He leads the attack, while the others follow and support him. When the combat is finished he returns to his former position, and the group resumes its patrol under the original chief. If a patrol meets two groups of enemy planes at the same time, it divides automatically, the left wing forming one group, the right wing another, and each goes to meet one of the Boche groups. When either terminates its combat it rejoins the other, and when the second combat is over the patrol re-forms and returns to its beat.

On the return of a group to its aerodrome, the chief of the patrol descends first, the others circling around to take their turns at landing, which they do in the same order as when they took their departure, at about two-minute intervals.

In flying over the lines one never holds to a straight course. This is not wholly due to the enemy's anti-aircraft guns, though it is a safeguard against them. It is also to give the pilot a chance to see in all directions so that he may not be surprised by an enemy plane. Remember, a pilot has six directions from which an attack may come— north, south, east and west, above and below. He must have a keen eye to be ever watchful in all those directions, lest death swoop upon him unawares. He must make sharp *virages*, banks and turns. He never looks over the edge of his *nacelle*, but always turns the plane on its side. Above all, he must never be overtaken by an enemy plane—he must overtake it. It is always every pilot's duty when in the air to overtake every plane that he sees and make sure of its identity.

The two-seaters usually go out with escorts, and in attacking them (they are usually of great value, since they carry observers, often with cameras) the first thing to be done is to separate them from their convoys. It is suicide for one man to attack a group, as the attacker would come under the crossfire of all their machine guns and would not have many seconds to live. There is almost complete safety in groups; and because of this pilots are forbidden to fly over the lines alone. Remember, the government's investment in a pilot usually exceeds five figures.

The departure at the termination of a combat—when you have decided that the chances are against you or you have given up hope of being able to outmanoeuvre your enemy—is most delicate. Often at this moment the observer of a two-seater will start his deadly fire g against you. It is now that the acrobatics are most essential, as by falling into nose spins, wing slides and the "falling leaf" drop you not only get away quickly but destroy your enemy's aim.

The nose spin, or *vrille*, has long been considered the best method of breaking away, and it is almost absolutely safe with an ordinary opponent. But should the Hun be an expert he will take advantage of the "dead points" in the spins—the moments when the plane's wing area is opposite him at the start and finish of each turn—to open fire. The wing slip is another excellent method of escape—it is used almost exclusively by Lieutenants Dorme and de la Tour, two of France's greatest pilots.

The laying of traps and ambuscades for enemy planes has become a recognized feature of aerial warfare. The object, of course, is always to bring a sudden and overwhelming force to bear against one or two men, shut off their escape and bring them down without exposing your own pilots to any serious danger. The plan most used was first tried out by Escadrille N-8—that famous organization in which Guynemer was the star. A patrol from it would go out, flying very high up, leaving an experienced man for a decoy 5,000 or 6,000 feet beneath them. They would keep out of sight as well as possible, while the decoy would loaf along, letting himself be "surprised" by a Hun. He would let the Boche come within 200 or 300 feet and open fire, then, before he got into any serious danger, go into a *vrille* or wing slip and escape. By this time the other members of his squadron would be on top of the Boche pumping shots into him from all directions.

The trick works both ways. It was in this manner that James Norman Hall, of the Lafayette Squadron[1], and the author of *Kitchener's Mob*, was so severely wounded in 1917.[2] His experience is almost incredible. The Lafayette was going on its regular patrol and Hall was delayed several minutes by motor trouble. This remedied, he hurried to the rendezvous; but his comrades had already given him up and

1. *The Story of the Lafeyettes-Escadrille*, by George Thenault, also published by Leonaur.
2. James Norman Hall was recently wounded in action and made prisoner, (as at time of first publication).

departed. He took the great risk of setting out to find them.

Five or six miles on the other side of the Hun lines he saw six or seven planes which, at their 16,000-foot elevation, looked like those of his patrol He started toward them and when almost beneath the group saw a single Boche plane below him. Still believing that the men above were his friends, he attacked the Boche. As he went into a *virage* to attack he saw that the higher planes were all diving for him, and that they were German! Instead of turning tail, Jimmy resolved to stick it out and dove straight at the Boche beneath, opening fire with his machine gun in the hope of getting his victim before he himself was brought down. Then he got it from every side. One ball passed through his left lung, another furrowed his forehead, a third pierced his thigh. He went into a *vrille* at full motor—a terrific manoeuvre —his only chance. He lost consciousness, and did not regain it until he had fallen to 6,000 feet—a 10,000-foot drop. Then he came to, just for a second, and he remembers distinctly bringing his plane back into control, shutting off the motor and heading for the French lines. He knew nothing more until he was on a stretcher, being carried to a dressing station;

A French officer who was in command of troops in the front line and witnessed the entire combat, completes the story. In a letter he wrote to Hall he says that after Jimmy came out of his *vrille* he saw him glide across the German lines into French territory, make a *virage* and redress his plane for the landing. Remember that all this time Hall was unconscious!. This surely goes far to prove the theory that an aviator's piloting is sub-conscious.

Nothing more amazing than Hall's landing has happened during the war. His plane came down directly over a French trench, the two wheels on its landing *chassis* set themselves down into the trench, and the machine was held firmly—not across the trench, but with the fuselage pointing straight up and the wings lying out on either side along the trench. The machine settled, the wings bent and collapsed, and Hall, still strapped to his seat, was let down easily into the trench, sustaining no additional injury from the landing.

It was afterward learned that the group which had attacked Hall was Baron von Richthoven's famous "flying circus," among the most dangerous of the Boche squadrons.

CHAPTER 5

More Work!

In the early days of the great war it was only occasionally that the shadow of an aeroplane flashed across the battlefield. Today there is no busy sector of the long battle line where a man with good glasses on any clear day cannot see half a hundred war birds soaring above him. And this is only the most obvious indication of how the shadow of the aeroplane has spread over all the operations of the struggle.

With thousands of the best brains of the world agonizing for any least advantage that may aid against their enemies, it was inevitable that they should seek every possible means of employing this newest engine of war—that each hoped, in some way or other, to out-think and so out-fight the foe with this untried weapon. So the aeroplane has come to be used in numberless ways of which the layman has only the faintest inkling—ways which have made it an assistant, at least, in every branch of the service.

Many of these ways cannot yet be told. There are some which we of the Allies still believe we have kept from the Hun. Possibly there are some, too, in which he has outguessed us. But even of the methods which are known and used by both sides, or could be used if the Germans had the hardihood which our men have shown, there are several which are almost unknown on this side of the water. Most people in America seem to have the idea that an aviator's only duties are to fight other aviators, to drop bombs and to do observation work. Their understanding of even these few duties seems to be very hazy, and there are many beliefs about them which seem queer to a pilot.

To every corps of the French army there is attached one *escadrille* which is not expected either to drop bombs or to fight—except in self-defence. This is the "*escadrille d'armée*," which works in connection with the Intelligence Bureau. This bureau does nothing but seek information about the enemy. Part of the work the *escadrille* does, of course, is to scout and take photographs. But—the Intelligence Bureau controls the spy system of France, and, make no mistake, the French (and the British) spying is as active and as successful as the German, even if it does not proclaim itself so. loudly. One of the hardest problems facing those in charge of a spy system is to get men through the enemy lines. Does this suggest any unadvertised activity by the aeroplanes?

It frequently happens that an aviator is called upon to undertake some mysterious duties outside his regular line of work, duties involving extra hazards and special knowledge, which are called "*missions extraordinaires*" Let me tell of one that came within my own knowledge—one that happened too long ago for the facts to hurt anyone now. The Germans are doing the same thing and know that we are doing it. The only thing is to catch each other at it. That sometimes happens.

Adjutant Batcheler of my *escadrille* had a call for 3:30 one morning. He was the pilot of our *biplace* (two-seater) Spad. Two other boys who were chasse pilots were also called for the same hour. None of the rest of us knew the reason for the call, nor the probable mission, though some of the older heads may have suspected. I was new, naturally curious, and so I asked the pilots who had been detailed to waken me when they went out. It was still dark, of course, with a few stars twinkling, when we left the barracks and walked across to the hangars. The *mechanicians* had the two-seater and the two *chasse* planes already wheeled out.

I was surprised to find my captain himself on hand. The pilots donned their fur lined suits and climbed into their seats. The *mechanicians* started the motors and let them run until they were warmed up. In the fitful lantern light I made out a Walloon peasant—one of the two types into which the Belgian peasants are divided. The French-speaking Flemish is the other type, while the Walloon speaks a Germanic dialect. This peasant was standing in the shadow of one of the hangars, dressed in the usual peasant costume, with wooden shoes, breeches, cap and jacket characteristic of the race. He was a hunch-back.

After testing out his machine, Batcheler called out, "*Ça va*" and my captain spoke to the peasant, who then came out of the shadow. I was struck by the fact that the captain's tone was more than kindly—it was respectful, almost deferential. This is not always the case when a

TWO-SEATER GERMAN RUMPLER, DRIVEN DOWN BEHIND THE FRENCH LINES

captain addresses a peasant. I was further astonished when the peasant climbed into the two-seater with Batcheler. There followed a short conversation in which the peasant seemed to take the lead, and finally almost to give orders. When the question at issue was decided the captain stepped over to the other two pilots, and gave them their instructions. The word was given, the machines started, rolled the length of the field, turned, and, just as the first streak of daylight appeared, thundered up into the air. They turned toward the lines, the escort trailing behind. In a minute they were gone.

The rather unusual terms of intimacy which exist between the officers and men of the *escadrilles* permitted me to ask my captain what was going on. He hesitated, then said, "They are going on *a mission extraordinaire.*" That was all, and I had, perforce, to be satisfied with it for the time.

I went back to the barracks, where I found the cook had a pot of coffee boiling, and sat down to a cup of it, still wondering. Finally, instead of going back to my bunk, I decided to wait till the machines returned and see if there was more to be learned. It was fully an hour and a half when they reappeared—remember, these machines can travel over 150 miles an hour—and as soon as the pilots returned to barracks I nailed them. I was all questions, but Batcheler, with cryptic smiles, was inclined to be reticent.

Finally, however, he told me the facts. They had taken the supposed peasant over the lines. The Boche spotted them as soon as they crossed, and opened an exceedingly brisk shrapnel fire. They ran on till they were some twenty kilometres back of the German front and over a field adjoining the forest of H———. With the shrapnel still bursting, Batcheler made his machine act as if it had been hit, and went into a *vrille* until he was only a little way above the ground. He straightened out and made a landing, but as the ground was rough and unfamiliar he ripped off his *béquille*—the protruding stick at the end of the fuselage, which, in conjunction with the wheels, helps to stop a machine after it hits the ground. Even before the plane stopped moving the peasant had jumped out and bolted into the brush.

Batcheler was up again instantly, at full speed. The others were waiting for him, guarding against any Boche planes that might be aloft. Then the three dashed for home through an exceedingly heavy shrapnel fire. All got through without injury, but when Batcheler tried to land, without his *béquille*, his machine *capotéd* (turned turtle) and was wrecked. Batcheler was, as usual in such cases, unhurt.

Later I learned that the apparent peasant was a man on whom Germany would gladly have laid hands. His costume, his dialect, all were assumed—including his hump. There was a very real reason for this deformity. The hump was a wicker basket, made to conform to his body and to give the outline of a real hump, but divided into four compartments, in each of which was a carrier pigeon. These pigeons were to carry back such information as he was able to obtain, but, be sure he used them only when his information was of great value, since when they were gone his mission was ended. Then he would have to make his way back into the French lines as best he could.

How many succeed in doing this, how many fail, only the chiefs know. I never saw nor heard of this man again, though I came to know much about the system. He is only one of thousands who deliberately risk their lives, knowing that detection and capture mean certain death—at best—and who return into the German lines again and again.

Of course, there is a reverse end to the traffic. Aeroplanes go out into the night with only a pilot and return with a passenger. Others, hovering far back of the enemy lines, perhaps over villages where there is hardly a sign of war, will drop swiftly, snatch a packet or a verbal message and wing back across the front. How many such trips are made I should not care to guess—information is very valuable. The identity, plans and movements of spies are concealed with the utmost care, even from those on whose side they are working.

Of course, the Allies have no monopoly on this line of work. The German spy system overlooks no chances. Whether the spy is a hero or villain depends on which side he is on. But of one thing we may be sure—not one of the Allied spies has resorted to the use of disease germs, setting fire to factories containing hard working women, distributing poisoned candy to children—or other dastardly methods common in the German system.

In passing, a tribute should be paid to the work the carrier pigeons have done in this war. They are the messengers of a thousand uses, and many thousands of them have been killed, since every soldier will, if possible, bring down any that he sees. If the bird's message is for his own general, it can be forwarded quickly—when it is for the Boche there has been a considerable gain made, for we may learn what his spies know and perhaps keep his commanders from learning it.

Portable dovecotes, built to house fifty or more birds and mounted on wheels, trail behind motor lorries which move them from sector to sector, keeping the supply at the front stations always full. The caretakers and trainers live in the lorry, and are always connected by telephone with the Bureau of Intelligence at the nearest headquarters. The birds are trained from these portable cotes, and almost always find their way back to their own home, no matter how far it may have been moved since they were taken from it. This is in spite of the fact that there are scores of cotes, all much alike. Sometimes they lose their way—perhaps from shell shock or some other incident of battle, though the turmoil never seems to frighten them—and in that case they will alight at another cote.

Perhaps in no branch of the service do the Germans make greater efforts to corrupt the Allied soldiers than in the aviation corps. An aviator has not only wide knowledge of affairs on his own side, he has unrivalled opportunities for delivering information safely across the lines, since a message can be dropped anywhere with absolute certainty that it will find its way to headquarters within a few minutes.

There was a case once. Three Allied aviators were returning from a patrol in the early morning, when they saw three planes coming from the German side and headed straight toward the Allied lines. The German anti-aircraft guns were not disturbing them; there could be no question that they were Boche. The Allied pilots "towered" to prepare for an attack. Then, just before the three Bodies reached our lines, the two outside planes turned back, while the third or centre one went on confidently. The Allied pilots swooped on him—and saw that his plane bore one of the Allies' markings. There was no trial. He fell blazing almost into the field of his own aerodrome. He paid the price of his treachery.

One of the most remarkable sources of information possessed by the Allied aviators was due to the ingenuity of Captain Georges Guynemer, the Ace of Aces. The ordinary Spad carries but one machine gun, mounted directly in the centre and in front, over the motor. Guynemer had two guns mounted on his personal machine, one on each side, thereby giving him double the number of rounds of ammunition carried by the ordinary pilot. But this was not his only reason for "packing two guns." It enabled him to mount, directly in the centre, where the single gun would have been, a telephoto camera, focussed along the line of his machine gun fire. The shutter was automatically geared to the triggers of his machine guns, on the control stick. A good pilot never fires until he is close, very close, to his enemy—with the fixed guns headed directly at the other plane. So when Guynemer opened fire he automatically took his enemy's photograph. A mechanical device changed the plates almost instantly, and sometimes he would bring back three or four exposures recording the stages of each of his battles. Sometimes these were so close that the very features of the Boche were discernible.

But more important for the Intelligence Bureau is the fact that these photographs give a complete picture of the Boche planes. Every type, in every position, is shown, and from these pictures silhouettes have been made and distributed along the entire front, giving the whole aviation service the latest information about any new Boche plane.

The work of the aeroplanes in co-operating with the artillery has been frequently commented on, but perhaps a few details of the way the work is done may still be of interest. The machines used for this work are two-seaters, and are often of an old, slow type, since they are not supposed to fight, but depend on the chaste patrol for protection against the Boche planes. They are equipped with a wireless outfit, the sending aerial trailing out behind. In addition to the pilot each carries an observer, a man of the highest intelligence and of long training in the delicate work he has to do—radio work, signalling of all kinds, photography and map reading.

Every salvo that is thrown by the battery to which the plane is attached must be reported—and remember that the plane must report shots from one particular battery when there may be hundreds of guns at work on every mile of the front under its observation. The target at which the battery is shooting must be known exactly, and the exact distance from the target as well as the direction of the spot at which each shot strikes must be reported. While the plane can report to the battery by wireless, very few machines have yet been equipped with a receiving apparatus, so the battery must communicate with the plane in regard to any change of target, etc., by means of what is known as the panel and shutter code. This is an elaborate signalling system, developed during this war, which locates targets by means of a series of numbers on the map which the observer carries with him. When he has centred the fire of one battery on the target assigned, the ground signals give him a new target and he begins work all over again on that.

These artillery regulating pilots and observers work in two-hour shifts, taking the air twice each day. They keep up their watch through every bombardment, stopping only on the appearance of their relief,

French Bi-motor Caudron ready to take the air

which is generally another plane from their own *escadrille*. They always know the pilots who are to relieve them and when they are due; when that time comes they signal down that they are going home. At this change of shifts many humorous *adieus* are sent down —some of them rather grim, for the boys will not take life and death very seriously. The observation planes do their work at low altitudes, from 6,000 to 7,000 feet, whereas the *chasse* pilots work above 10,000 feet and often as high as 20,000.

The present organization of the artillery in use in this war is as follows. There are three brigades of artillery to each division and two or more divisions to each army corps. Each brigade consists of three batteries of guns or howitzers. A battery of heavy artillery consists of six 5-inch guns; a siege battery consists of four 10-inch howitzers.

In general terms, there is this difference between a gun and a howitzer, as the names are applied at the front. A "gun" is a piece of artillery having a low elevation, or comparatively flat trajectory, but great range and searching power. A howitzer has a low muzzle velocity but a high elevation, about 80 degrees, so that its trajectory is very much more curved; it has not the range of a gun but is much more accurate and is used in shelling trenches, firing over hills, into valleys, etc. Siege batteries usually fire heavy explosives while most of the shrapnel is fired by the guns. In addition to the heavy artillery and siege batteries there is the super-heavy artillery consisting, generally, of long range naval guns mounted on flat cars which run back and forth on a track laid along the lines.

At the beginning of operations in a new sector the entire enemy country is photographed from the air. Every gun position is noted, every railroad siding and spur, in fact everything which it is desired to destroy is identified and located by these photographs; then, by means of these photographs, these points are marked on a new map made for that particular purpose. This map of a sector is then divided and subdivided into squares and divisions as follows:

First, the entire sector is divided into twenty- four equal sections, each section being a square marked by a capital letter. There are four horizontal rows, six sections to a row, and they are lettered from A to X inclusive, from left to right. The first subdivision of these sections is into 80 squares for the top row of six squares (A to F inclusive) and also the bottom row of six squares (S to X inclusive); and into 86 squares for the two middle rows (G to R inclusive) . Each of these first subdivisions, Al, A2, A3, etc., is again divided into four squares, each quarter being designated by the small letters, *a*, *b*, *c*, and *d*. Each of these *a*, *b*, *c*, and *d* squares is further divided into 24 numbered squares, these last being the smallest subdivisions.

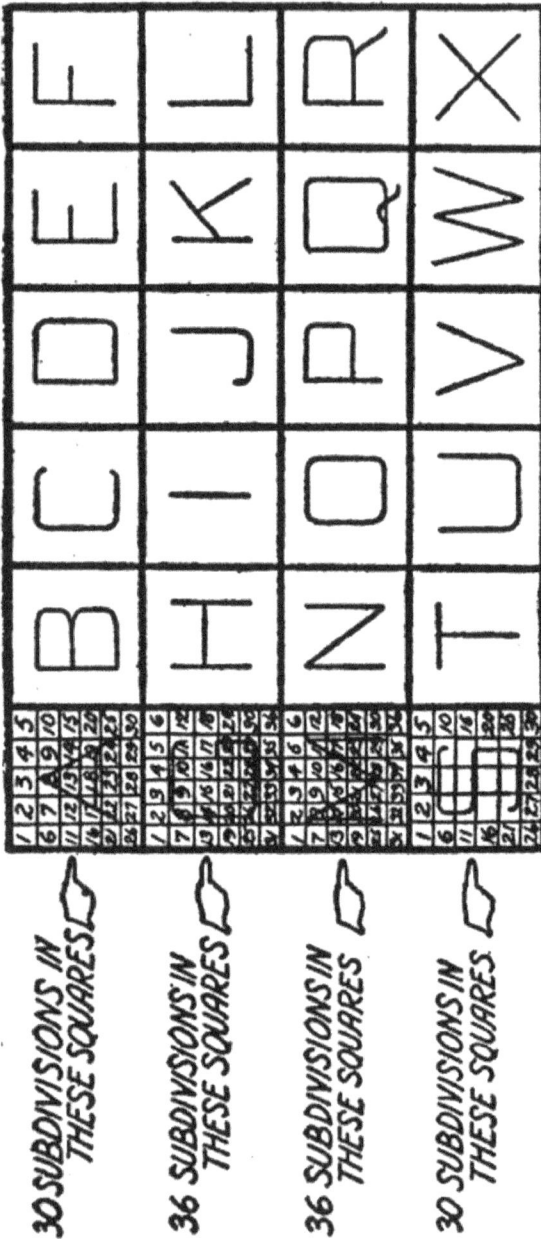

FIG. 1. DIVISIONS AND SUBDIVISIONS OF A SECTOR

The first division of a sector into *sections*, lettered A, B, C, etc., and the first subdivision of these sections into numbered squares, as shown in Sections A, G, etc.

Therefore the location of some object or point, which might be designated for attack or destruction, could be indicated, for example, as being in square A-7-b-16. Perhaps an enemy battery or ammunition dump, which has been located and spotted on the map, is to be the target; the location is specified by letters and numbers as indicated. As there are 2,880 or 3,456 of the smallest squares in each lettered section (A, B, C, etc., or G, H, I, etc.), according to the letter designating the section, or a total of 76,032 squares in a map covering an entire sector, each square represents a comparatively small space of ground. Therefore, a target can be located with almost minute accuracy.

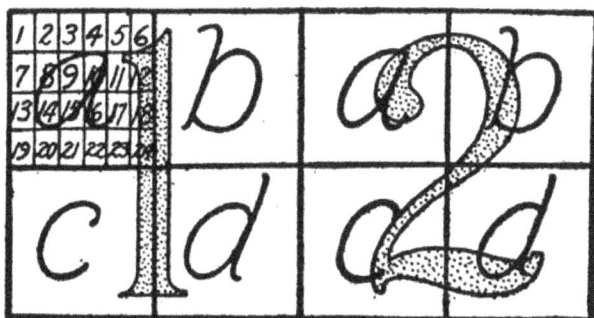

FIG. 2. THE SECOND, THIRD AND FOURTH SUBDIVISIONS OF A SECTOR, DRAWN ON A LARGER SCALE.
The large squares, numbered 1 and 2, and divided into quarters, a, b, c, d, correspond to A-1 and A-2 in Figure 1. The small squares, numbered 1 to 24, indicate the smallest subdivision.

Squadrons of regulating planes are attached to certain brigades of artillery and detailed to work continuously in conjunction with them. A pilot going out on artillery regulating service first ascertains from the brigade commander by telephone what the first target is to be, the location of the battery working on it, the kind of battery, the type of explosive to be used and the duration of the fire. Then, if possible, he obtains a photograph of the target for better identification, consults with his observer and, when everything is understood and ready, he takes to the air. First, he circles over his own aerodrome to test his radio set, then if everything is working right he flies to the battery whose fire he is to regulate. Arriving there, he radios his presence to the battery, and when they reply with ground signals, the work begins.

One method of correction of fire is by what is called "The Clock System," which, undoubtedly, is very efficient. The observer, seeing where each shell hits the ground, communicates to the battery, by means of his radio, the location of the point where the shell strikes, relative to the target, so that the gun pointer may correct his aim accordingly for the next shot. The signal consists of a figure and letter, the *figure* corresponding to one of those on the dial of a watch and indicates north, east, south or west, or the intermediate points of the compass, relative to the target, the target being represented by the centre of the dial. The *letter* represents distance from the target, indicated by imaginary circles drawn around the target as a centre; each letter being equal to an arbitrary number of meters and measuring the radius of the imaginary, concentric circles. Figures 3 and 4 are a graphic illustration of this system.

For example, 4-C would mean that the shell hit east by south, 200 meters short of the target; 9-A would mean, west of centre 50 meters. In order to avoid confusion of signals, for there are numerous planes doing the same kind of work at the same time, each pilot, each *escadrille*, and each battery has a code number or letter. By using these codes in sending signals each receiving operator identifies the signals intended for him.

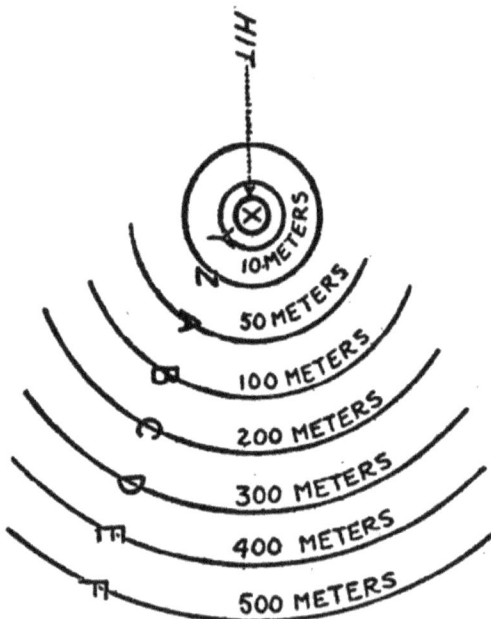

FIG. 3. SHOWING THE MANNER IN WHICH DISTANCES FROM A TARGET ARE INDICATED BY LETTERS, IN ARTILLERY-REGULATING BY AEROPLANES.

On the Meuse, in the early part of 1917, the Boche succeeded, through spies, in obtaining the battery number of one of our 75s. They immediately sent one of their regulating planes into the air which got in time with the antennas of that battery and misdirected its fire so that it began shelling our own men. Before the treachery was discovered hundreds of our brave fellows were mowed down. Of course, it is possible for both sides to work that trick, but for that reason the code and the method of plotting are being changed constantly; so I am giving no information that the enemy does not already possess or that could be used against us.

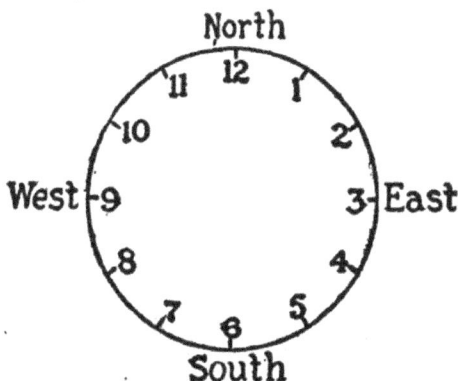

FIG. 4. THE POINTS OF THE COMPASS, AS INDICATED BY A WATCH DIAL.

Incidentally, one of the greatest air dangers comes, not from enemy planes or from anti-aircraft guns—the *chasse* planes are far beyond the reach of these little beasts—but from the high angle shells of their own and enemy guns fired at other targets, often far behind the lines. The howitzer shells often reach a height of miles at the top of their trajectory, and the fighting, as well as the regulating planes, may be in their paths. It is a curious thing that pilots are very often able to see these shells as they pass by. I myself have often seen them clearly enough to recognize their size and shape.

One of the most remarkable things that has happened at the front was the clipping off of the upper wings of a Caudron type of plane by a German shell. The plane was in the act of turning; it was sidewise to the lines when a shell passed right through the struts and took both upper wings off clean. Stranger still, the pilot was able to land back of our lines without either the observer or himself being seriously injured. They had a bad shaking up, of course, and smashed the plane, but they themselves were only a bit cut up.

CHAPTER 6

The Eye Above the Battle

Dipping, soaring, climbing—never for a minute taking a straight course and never for an instant breaking its alignment—a group of aeroplanes sails like a flock of tumbler pigeons above the reek of No Man's Land, and is now lost, then seen, behind the black shrapnel puffs of the Boche. Straight into the heart of it they fly—away from France and into the domain of the enemy, a veritable maelstrom of flying steel.

It travels in a V-shape formation, point foremost, the largest of the eight machines, a two-seater, at the apex. Three hundred meters to the right rear and 300 meters above the leader is a smaller plane; 300 meters further to the right rear and on the same level is another. Two similar flankers form the other leg of the V. Directly behind the leading plane, but 300 meters above it, is another, and at a distance of 500 meters from a line joining the two outermost flankers and about 500 meters apart come the trailers of the flock—the hardiest fliers and the bitterest fighters (*Les As*—The Aces—of the *Escadrille*) .

This group of fliers, always swooping, circling, dodging among the bursting shrapnel and always maintaining its compact formation, is one of the latest developments in aeroplane scouting. It is the most trying work in all the air service, the most productive of military results and the least productive of glory.

Within an hour after it returns General Headquarters will have a new series of photographs showing in minute detail the situation behind the German lines. Five minutes later shells may be falling on some important point revealed by these new pictures.

Fifteen or twenty kilometres behind the German trenches the aeroplanes go, to some predetermined objective. Here the leading plane will swoop down, probably to an altitude of about 4,000 meters, while its companions—or escort—circle in a kind of mad but orderly pin-wheel dance above it. In a few minutes a dozen plates on each of three cameras attached to this plane will have snatched a map of the countryside below; the big plane rises, the smaller ones dart to their accustomed places and the perilous flight, amid incessant puffs of shrapnel, back to the French lines begins.

Such a group is an *escadrille d'armée*; one of which is attached to each Army Corps in France. It is a subdivision of the first of three branches into which the department of aviation is divided. The three branches are as follows:

The *avions de chasse*, or pursuit planes, are single-seater machines, each carrying a pilot. These are the scouts and fighters.

The *avions de régalement*, or regulating planes, are two-seaters, carrying a pilot and an observer. Their function is to regulate artillery fire by "spotting" shots and correcting ranges by wireless.

The *avians de bombardment*, or bombing planes, are the night fliers that bomb important military positions.

The *avians de chasse* make up the *groupes de combat*, each group consisting of four *escadrilles* of fifteen pilots together with an *escadrille d'armée* for each Army Corps.

Although an *escadrille d'armée* belongs to the *avion de chasse* branch, it contains one two-seater plane. This two-seater is the eye that scans the prospective battlefield from the clouds, and it always works with a protective escort of fighters. The two-seater is more unwieldy and less adapted to the tactics of air combat than the swallow-like single-seaters, but it is the very "apple of the eye" of its fighting convoy. Affixed to its fuselage are three downward pointing cameras, the last word in telephoto construction. They are so arranged that the field of vision of each ends precisely where that of its neighbour begins, yielding a triple panoramic view of the country far below the photographer's feet. Their manipulation is simplicity itself. The pressing of a button blinks the shutters and a turn of a lever shifts the plates in all three cameras.

Two sizes of plates are exposed, 4"x5" and 6½"x8½". Two kinds of lenses of different focal lengths are used, one with a focal length of eight inches and the other of ten inches. The field of vision of course depends upon the altitude at which the pictures are taken. The eight-inch lens, at an altitude of 8,000 feet, will photograph 2,222,272

AERIAL PHOTOGRAPH OF GERMAN AERODROME

square yards of ground; at a height of 10,000 feet the same lens will cover 3,422,222 square yards.

A glass panel just below the photographer's feet is his finder and indicates exactly the limits of his field. It is astonishing to know at what heights photographs clear enough for use in military maps can be taken under exceptional conditions. I have seen an excellent photograph, with minute details discernible to the naked eye, which was taken from a height of 17,000 feet—about three miles and a quarter.

The photographer learns to estimate the altitude at which weather conditions will permit him to use his cameras. The various *strata* of mists, the position of clouds relative to the sun, their size and colour, all come under his practised eye before this group leaves on its perilous mission. And they must always fly high, where there will be less chance of shrapnel interruption or lurking Boche patrols.

As the result of constant work at this business of photographic map-making strange atmospheric freaks and their effects on photographic plates have been noted, and gradually, as the flying photographers compare experiences, the science of air photography is being developed. Generally speaking, an absolutely clear sky affords the best possible conditions. A high cloud layer that reflects light down to the earth also results in good pictures. Clouds that are broken and lower afford an opportunity for concealment until the actual moment for shutter-snapping arrives.

Massed, low-hanging clouds, rain or thick mist usually make photography impossible. It has been learned, however, that a certain form of mist, almost impenetrable to the human eye, offers no difficulty to the fast lens, but in fact seems to filter the light rays in such a way as to produce an extraordinarily clear picture. So far as I know no explanation of this phenomenon has been found. Military photographers, like military aviators, are not inclined to worry much about causes unless they underlie conditions that must be remedied. Conditions that fight for us are accepted with gratitude.

Two and sometimes three times a day the group goes out, flying perhaps fifteen or twenty kilometres behind the German lines, running a gauntlet of continuous shrapnel fire from the anti-aircraft batteries every foot of the way—all to win a two-minute opportunity to hover above some point that is not yet a part of the great new photographic map the experts are patching together back at General Headquarters. Sometimes they are in the air six or seven hours in one day. New maps are being made constantly, especially before an attack.

Always our chief duty is to get the two-seater plane back with the pictures. There must be no yielding to the lure to stop and engage in a duel with the Boche. We must pass him by. The pictures are our first consideration. It is tantalizing sometimes to see a Fritz sailing along below you, when you could hop on his tail and riddle him with your Vickers.

But it can't be done, for ten chances to one the silly-seeming Fritz knows just as much what he's about as the ruffed grouse that, with trailing wings, plunges squealing across your path when her brood is close by. A more sinister motive may impel the apparently reckless Fritz. Ambushed behind some cloud bank are probably three or four more of his kind, waiting for the chance to assail the two-seater if we are lured into the trap and leave the photographer without full protection, so that the odds would be in Fritz's favour. That chance might come if our formation were broken by one or more machines swerving off in the hope of adding another Boche scalp to their string. Those pictures are worth more than a dozen German planes. They *must* be brought back. Every personal consideration, meaning army citations and palms, must be sacrificed to the execution of this military mission.

Moreover, the plates are not the only treasures that we are convoying. The machine that carries them is one in itself. The two-seaters used on an expedition of this sort are the much discussed "Spads," with the Hispania-Suissa motors of 200 horsepower. Ordinarily their speed is 120 miles an hour, but they can do 140 if crowded. But there are certain wrinkles about the picture-taking machine that are brand new. None of the late model two-seaters has thus far fallen into the hands of the Boches, so they don't know all about our system of mapping yet. And we, the fighting escort, must see that they don't.

Members of these convoys have been especially trained in group flying and tactics are prescribed for almost every conceivable situation. There is scarcely an attack possible that will not set in motion a series of "plays" or manoeuvres, much in the fashion that a football team will run off a series of plays on a single set of signals. Of course, all these evolutions have for their object the protection of the plates and the machine carrying them.

If the Boche patrols fail to see us, the black puffs of their anti-aircraft guns soon tell them of our presence and they aim to cut us off from our lines. But we must return with our plates; we must dodge, however much we might prefer to fight it out with Fritz. Often over our own lines our first intimation of the presence of an enemy plane is the white puffs of the Allied shrapnel seeking him out. The Boches use black bursting charges, while ours are white. White puffs, near home, are danger signals, they mean that enemy air craft are over our lines.

The Boche seldom attacks us unless he has superior numbers on his side. We know that, so we let him show off his whole bag of tricks in an endeavour to entice us into combat: He has some pretty fair tricks, too. In case the Boche is waiting with superior numbers and actually does assail one of our planes, the squadron is arranged to afford the best possible defence.

An attack on the first plane in the V would bring the second machines and the one in the centre to the rescue. An attack on one of the second planes would bring the centre man and perhaps one of the trailers into the fight. The formation was adopted for this very purpose. If, however, the challenge developed into an air engagement which drew off two, three or even four of our planes (and it would take more Boche fliers than they can often spare to accomplish this) the remainder of the convoy would adopt a prearranged formation and pursue its course homeward. If one of the trailers is attacked he fights it out by himself, or, at most, with the assistance of his companion trailer. The rest go on without them.

Although the machines are uniform in appearance and colour, it is easy enough to pick out individuals once you know them. No two pilots fly exactly the same. They have peculiarities, just as men have in their walking. There is, of course, no means of communication between the planes while in flight, except by the execution of certain manoeuvres. Each pilot will have some identifying mark painted on his machine, like the war paint on an Indian's face, or it may be that he has painted his own or some dear one's initials on it.

As soon as we reach our own aerodrome, the formation is broken, and we make our landings. The chief of the laboratories is waiting. He and his assistants rush to the observer's machine, and, almost before it has stopped rolling, dismount the cameras, and the plates are on their way to the developing rooms. These are in *camions* (motor trucks) fitted up with dark rooms, tanks, drying racks, and all the newest photographic appliances.

The plates are developed and then dried by a chemical process so swiftly that prints can be made and sent to headquarters within an hour after the return of the two-seater. Five experts are at work constantly, reducing these pictures to scale, determining the overlapping lines and pasting together a marvellous photographic map. This work continues until the entire sector has been mapped.

In the meanwhile, a different type of work is going on. Duplicates of the photographs have been rushed to the General Staff Office, where artillery observers with magnifying glasses search them carefully for indications of new works of military importance—ammunition dumps, communicating railroads, fortifications, troop concentrations, or anything else that should receive the attention of our heavy guns. The moment such a place is located its position is flashed to artillery headquarters, and in a few seconds shells are dropping on the spot.

Once having viewed these maps, it is easy to see that they are the only ones of real military value in preparing to make an attack or to resist one. Every road and every shell-hole on it, every house, every barracks, every ammunition dump, every railroad, every siding, every depot is shown in detail. Scouting trips now and then, with a few pictures snapped here and there, serve to keep the great map correct to within a period of hours.

For these corrections sometimes a smaller squadron is used, consisting only of three or four planes. To these the Boche pay little attention, thinking the group merely a patrol. The observation plane with these small groups is a single-seater like its guards, and as handy to manage. It is equipped with only one camera; underneath and to the left of the pilot is a reflecting mirror which brings the camera's field of vision to within a few feet of his eyes. The button and lever which work the shutter and change the plates are on the driving control of his machine, so that he scarcely has to move his hand to reach them.

The single-seater Spads carry a built-in machine gun of the Vickers type, which is pointed by aiming the plane. The two-seaters have one stationary machine gun in front and another which the observer operates mounted on a swivel in the rear.

There is other work, too. Twice a week a little job of newspaper delivery must be done by the *Escadrilles des Armées*. Little fruit has fallen as yet as a result of this work, or if it has it hasn't made much of a thump, but the delivery goes on with the utmost regularity and with the greatest faith that persistence will win.

Literally, the French fliers are the paper carriers for the Boche. "Tracts," we call the papers. They are newspapers, printed in German of course, supported by German-Americans and democratic Germans in neutral countries, which are designed to carry the gospel of democracy to the Boche. They are the *Neue Deutscher Freie Presse* and the *Frankfurter Zeitung*.

Each plane takes about 1,000 "tracts" aboard and goes sailing off on its missionary errand. The objective of each is some barracks where it is known that considerable numbers of soldiers are quartered. The "tracts" give the news of conditions in America, of the nations that have severed diplomatic relations with Germany, and of the aims of the Allies. The entry of America into the war on the side of the Allies, with the promise of active military support, the news of the great war preparations under way in this country—all were told in the "tracts" which we showered down on the Huns.

Today the Boche is kept in touch with the progress of the Liberty Loan, and the purposes to which the money is to be devoted, by his faithful paper carriers. When the carriers arrive over a hostile barracks the soldiers come scurrying out by scores and hundreds to pick up the papers which go fluttering down. They know well enough when their semi-weekly sheet is due, and display not the slightest fear that the "newsboys" may be a bombing party. But they are chary about being caught with a copy in their possession, as they are severely punished for reading this literature.

And what a reception the Boche gunners give us "newsboys." The great burst of paper from the sky is the signal for an answering burst of shrapnel from the anti-aircraft guns all along the line. They display more animosity toward the paper carriers than toward any other invaders of their atmosphere.

I brought back several copies of the various issues of both these papers, and on letting some Americans of German birth read them, they unanimously declared them to be eminently fair and devoid of the malice found in the propaganda of our enemies.

Another class of work, and probably the most dangerous and the most exciting of all the tasks that are assigned to the aviators, is known as "*liaison d'infanterie*." This work consists in co-operating with the ground forces during an attack and takes two forms.

The first form is worked in this manner. When an attack is in progress, every *poilu* wears a white tin disk, twelve inches in diameter, strapped on his knapsack. During the advance the infantry must drop at intervals for rest and cover. When they lie prone in the shell holes and craters these tin disks, winking up at the observer in the air above them, make a white polka dot pattern on the landscape, and the observer can determine the exact distance the infantry has progressed.

This he wires back to the battery commanders, who extend their range so that the barrage fire is kept at a proper distance in advance of their troops. A miscalculation of the distance by the aerial observer would result in our men being crushed under the fire of their own guns. For this work the observer must fly low, sometimes not more than 700 or 800 feet above the ground, so that he may accurately judge the distance advanced. Armoured aeroplanes, the *nacelle* of which is covered with sheet steel, are now being used for this class of work because of the intensity of the fire from the Boche machine guns in the trenches opposite them. Heretofore planes used in this work were good only for the scrap heap after one flight, they would be so riddled by the enemy's bullets. The Germans have attempted to develop an armour piercing bullet for use against these armoured planes, as well as against the tanks, but so far their success has not been great.

The other form of *liaison d'infanterie* work is even more exciting, though probably less important. This work is assigned to the *chasse escadrilles* and consists in participating in the actual attack on the Huns. The damage done is perhaps not so great in the number of men killed as it is in the confusion of the Boche and the injury to his morale and at the same time in the encouragement it gives our own troops. In this work the pilots will often be from six to ten kilometres back of the enemy line, attacking the reserves who are frequently not under trench shelter. They especially delight to dive with machine guns popping against any bodies of troops they can catch in the open.

Chasse escadrilles are also assigned to attack certain aerodromes back of the enemy's front, to demoralize them and prevent their planes from getting into the air. So it is that on the morning of an attack each *escadrille* has a certain ground to cover, certain men to fight, certain ends to attain. One can imagine the excitement and confusion of the enemy who must contend with these aerial dangers as well as with the attack on the ground. It is a form of warfare which the German has so far not attempted, except in isolated instances.

The biggest game that the pilot hunts is the "sausage balloon." Along the western front is a continuous double line of these balloons, stretching from the Channel to Switzerland. They are used for observation of the enemy's first, second and third line trenches and are sent up from one to three miles back of the front trenches on each side. In the *nacelle*, or basket, of the balloon—which, by the way, in the later types is no longer a wicker basket, but a steel car, equipped much like the radio room on a battleship—axe two observers with high-power field glasses. Their distinguishing fitness for their job is that they are experienced parachute jumpers.

Each man always works with a parachute strapped to his body, ready to jump instantly if the balloon is attacked, hit by a shell, catches fire, breaks loose from its anchor, or is in any other serious danger. In any of these events the observers do not wait to descend with the balloon, but immediately "abandon ship" and get out of danger as quickly as possible by making their "leap for life," depending on the parachutes to land them safely within their own lines. There is a good reason for this aside from the element of personal safety, and that is the preservation of their notes and data, which they carry with them when they make the jump and which might be destroyed or fall into the hands of the enemy through the wrecking of the balloon.

Nowadays the balloon is anchored to an electric windlass on the ground, so it can be pulled down readily in case of an attack by Hun planes or if the Hun artillery finds its range. The bags are never deflated when they are lowered, but are pulled down by this windlass and allowed to rise again as soon as the danger is over. They generally have 4,000 or 5,000 feet of cable, and rise very quickly when they are let out.

The observation balloons are the first object of an attack whenever an advance movement is contemplated, or in fact when any other important manoeuvre is about to begin, as their destruction goes a long way toward blinding the enemy and destroying the power of his artillery. In 1916, when the great Somme attack was about to begin, a concerted drive was made by the Allied planes from the Channel to Switzerland against the observation balloons on the German side. It has been reported that, while they did not get everyone, they did succeed in keeping them out of the air for nearly three days, thus greatly increasing the chance of the surprise which was contemplated by the Allied Commanders.

An observation balloon is most difficult to attack because of the protection afforded by the anti-aircraft guns. These are carefully trained to cover the air immediately around the whole balloon. The reception the aviator gets when venturing too near a "sausage" can best be imagined. In addition the big bags are built in compartments, which the ordinary bullet merely pierces and does not destroy, and even the destruction of two or three compartments will not bring down the balloon.

Therefore, aviators attacking them always use incendiary explosive bullets, hoping to set fire to and explode the gas bag. In ordinary aerial warfare our machine gun belts are loaded with one so-called "tracer," or luminous, bullet to each ordinary shot, so that the little trail of fire will enable us to tell where our bullets are landing, but for use against gas bags the incendiary explosive bullets are substituted, and these are so delicately made that they will explode when they strike the silken cloth and rubber coverings of the bag.

One of the most amusing instances I ever saw on the western front was an attack by two German scouts on a Belgian "sausage." This balloon, instead of being anchored to a stationary winch, had a motor truck equipped with a windlass for its anchor. The anti-air-craft guns suddenly announced the presence of enemy planes by their terrific cannonade. The Boche, however, came through the fire, diving straight at the balloon from above. They got so near that, for fear of hitting the balloon and observers, the anti-aircraft batteries had to cease firing. Immediately the two observers made their "leap for life" and seemed to fall hundreds of feet before their parachutes opened.

They then went swinging through the air, their legs dangling and kicking most ungracefully. The motor truck started up the road at full speed, with the balloon trailing behind it, and at the same time the windlass began to wind in wildly on its tether. The balloon was actually pulled "out from under" the Boche and left them open targets for the anti-aircraft battery. This immediately resumed fire and brought one of the Boche down. This means of controlling and moving balloons is being developed rapidly.

One more aerial activity which can be discussed is bomb dropping. We hear a great deal more about the German attacks than those carried out by the Allies, because the Germans have the habit of attacking towns and killing women and children. They seem to prefer this rather than attacking railroad centres, ammunition dumps and other military objects, which are the only targets sought by Allied airmen, except when they are engaged in reprisals. The French have been much more free to adopt retaliation than have the British, and as a result their unfortified towns are now suffering very little.

In the early days of bombing it was customary for the pilot of a bomb-dropping plane to remain at a considerable height (6,000 or 8,000 feet), while getting rid of his dangerous load, but we have found that bombs dropped from this height accomplish very little. Of course, for killing women and children that height is all right, but a man who has a definite military object in view, such as destroying a railroad train or switch yards, must get much closer to his target if he wants to make sure of a hit. Allied airmen will now get within 500 or 600 feet of the ground, risking the fire of the anti-aircraft guns, in order to get the utmost good out of their loads of destruction. I think I may safely say that they do.

It must be remembered that a bomb has no sympathies. The pilot and bombardier have more than the attack of the enemy with his shrapnel fire to contend with. The knowledge that their cargo will blow them to fragments if their plane should catch fire or fall to the earth gives them plenty to worry about while on their perilous mission. This load of violent death must be delivered, too, before the plane can attempt a landing.

I remember one case when a huge Voisin bombing plane had gone out to amuse a certain Boche military position in Belgium. It carried its bombs on a mechanical contrivance, six under the lower plane of each wing, making twelve in all. On its return above its aerodrome the bombardier discovered that two of the bombs under the left wing had failed to fall on the release of the mechanism. To land with these bombs intact would be to invite sudden death.

So the machine flew out over the sea, while the bombardier strove in vain to release the bombs. For over an hour—the longest hour, no doubt, in both men's lives—they hovered over the Channel. Finally the gasometer showed that they had used their entire supply of fuel. They had to come down, and, strange to say, they did so without injury. They are still wondering whether it would have done the Boche any damage if they had succeeded in releasing those two bombs.

It is the *staccato* beat of the night bomber's motor that betrays him to the "bombee." All of these night hawks carry red and green signal lights underneath their wings and have to make the proper response to a prearranged set of signals (pass words) to get past our lines without arousing the anti-aircraft guns. But despite that precaution and the terrific uproar of the present motor the enemy gets over now and then—quite frequently, in fact.

On the Flanders front our camp was near that of some British officers and on the evening before my departure for America I went over to have a farewell dinner with them. We were seated in a marquee tent about 10 o'clock in the evening when there came the deafening roar of the anti-aircraft batteries on every side, and the spiteful chatter of machine guns. Through all the uproar we could hear the beat of a Boche motor sounding like a Fiat at the Grand Prix.

Somebody "doused" the lights and we went out. Such a scene as met our view was undreamed of and impossible before the war. The whole sky was alight with electric beams from horizon to horizon. Away off on the Channel patrol boats were fingering the darkness and every searchlight for miles behind the lines was weaving back and forth, sweeping gigantic circles in the darkness—searching—seeking.

Through the coldly luminous darkness that veiled the earth and the brilliant glare that lit the sky came bursts of flame as the anti-aircraft shells exploded by dozens, scores, then hundreds. A million criss-cross jets of fire marked the course of the luminous bullets from the machine guns that went streaking across the whole scene.

Shrapnel from our own guns was raining down on us thick and fast, so we put on our steel helmets in a hurry. Still, above the hideous din of the snickering machine guns and the clamorous anti-aircraft guns, which was punctuated by the thudding boom of the bombs the Boche was dropping, came the insistent drumming of the motor.

At last a searchlight found him and by a thrice repeated sweep revealed the target to the others. At once all the beams converged upon one point in the heavens and just for a moment we caught a glimpse of the Boche. He wasn't more than 4,000 feet up, but in an instant he was lost again. Only his noise remained.

Recollection of that scene and of the final escape of the Boche, despite the handicap of his motor, impels the most profound respect for a motor that would be noiseless.

CHAPTER 7

Famous Fliers

At the close of 1917 the *Department Aeronautique* of the French Army issued a bulletin, of which the following is a part:

Liste des As, au 1ᵉʳ Janvier, 1918.

Lieut. Nungesser.	30	*appareils* (Planes)
Capt. Heurtaux	21	"
" Deullin	19	"
Sous Lieut. Fonck	19	"
" " Madon	19	"
" " Lufberry	17	"
" " Navarre	12	"

Disparus (Missing)

Capt. Guynemer	53	*appareils*
Sous Lieut. Donne	23	"
Capt. Matton	9	"
Serg. de Terline	6	"

The list is too long to go over in detail, though much could be written of any one of these men. Some of the greatest are dead and the world will never know the details of all of their exploits for that reason. Even all that is a matter of record may not be known until after the war is over. How many of those now living will be left when that happy time comes? All of them, let us fervently hope. Yet all of them almost every day, weather permitting, fly into the jaws of death and no one knows at what moment fate may overtake the best of them; for even the best may not always avoid the many, many adverse chances that beset them over the enemy's lines. Their ability is unquestioned and unquestionable so far as holding their own, and more, against any pilots in the world today is concerned.

Yet they are but human and even a novice may accidentally get in a telling shot when fighting the best of Aces. A pilot's career is a game of skill but it is also a game of chance. The more skilful man *usually* wins but not always, when luck or chance plays against him. No man can foresee or forestall all of the exigencies of war when he fares forth to do battle. Achilles had his vulnerable spot; Roland and Oliver were valiant knights, yet each met his fate on the field of honour. War is war and even the most invaluable must sometimes be sacrificed to the insatiable appetite of the horrible beast.

These Aces[1] are, for the most part, modest men. They take their work seriously but, like most men who are truly great in any line of endeavour, they do not take themselves overly seriously. Therefore they are not given to talking so very much about themselves and their adventures. A hair-breadth escape or a skilful thrust that brings down an enemy is all part of the day's work and is taken as a matter of course with them. A man who constantly flirts with death is not prone to *braggadocio*, he *isn't that kind of man*; the loud talking braggart is always found in the safer and pleasanter places where his precious hide is secure and his tongue has opportunity to keep pace with his imagination as it wanders far afield. The garrulous kind never have been and never will be Aces, for Aces are not made from that sort of stuff. Hence it is not easy to get much information, first hand, of a personal or intimate nature from these men, even from those who still could talk if they would.

Guynemer was the greatest of all Aces, Allied or enemy. The world has not yet produced his equal and it may be many a day before it does. Yet it was undoubtedly by a mere chance shot at the hands of a novice, one of those little and sudden whims of fate that sometimes overtake the mightiest, that laid him low so that now he sleeps in a grave unknown to his friends somewhere back of the German trenches. His name is a household word in France. It is likewise known to almost every red-blooded boy in America who loves chivalrous deeds and high adventure.

The simple eloquence of Guynemer's last citation from the Commandant of the French Armies is as remarkable as it is brief. It reads:

Capt. Georges Guynemer, 54 Aeroplanes, 215 Combats, 2 wounds.

---

1. *Heroes of Aviation* by Laurence la Tourette Driggs, republished by Leonaur as *The High Aces.*

That is all. Words had failed them, superlatives had been exhausted, appreciation was beyond the power of language to give adequate expression. His record had been so marvellous, his previous citations so filled with words of praise and tribute to his valour, that at last there were no words left to convey their thoughts other than a recapitulation of his record. But after all, what commentary is more to be desired; what greater monument *could* be erected to any man than this one, builded in the hearts and memory of his countrymen by his own hands? Such, I believe, is all that he would desire, for of such simple and modest nature was the spirit and heart of Guynemer, Second Lieutenant Nungesser, the present-day Ace of Aces in the French Army, climbs into his plane at nine in the morning and remains over the lines until three in the afternoon, making but two landings during that time at some aerodrome nearer to the lines than his own to replenish his gasoline tanks and ammunition belts, so as to lose as little time as possible. Six hours in one continuous stretch in the air, always hunting, forever hunted; constantly on the alert, watching, waiting, dashing at any enemy who may venture within range of his vision, challenging, fighting, crushing the enemies of France and Liberty. One example of his results was—two Boche planes and one balloon in a single day's work of six hours. It may be apropos to add that six hours over the trenches is, on the average, worse than sixty in them.

Donne—"Pere Dorme," he was called in loving affection by the comrades who knew him as friend and brother pilot—was a small, rotund, mild-mannered and quiet-speaking man. Without his uniform, he had the air and appearance of a village priest. But clad in his pilot's combination suit and seated in his Spad, he played with the Boche as a cat plays with a mouse, not in a cruel sense, for he could never be of that nature, but in quickness, strength and sheer ability. Never hesitating to attack, no matter what the odds, he rode the air like a thunderbolt, leaving wreck and ruin among those who ventured to oppose him. His name is revered by all in France and when, some day, a statue is erected to his memory it will, if it is true to his life, reflect a personification of courage which, in such large measure, was shown in the career of good "Pere Dorme."

Sergeant de Terline, who gave single-handed combat to five enemy planes and, bringing one crashing down to earth, put the others to rout. Not satisfied with that, sufficient glory though it was for any man in one day's work, he followed the retreating enemy group. Catching them with his faster plane he gave further combat, was wounded, and then his machine gun jammed and he was powerless to send any more of his well-aimed shots after the enemy. Still undaunted, and unwilling that all of the prey should escape, he precipitated his swift plane into that of the nearest Boche and dragged him down to earth. When the two planes struck the ground both were in flames and both pilots had gone to meet their Maker. It is this sort of men that has secured for the Allies the ascendancy in the air and the name of de Terline will live long in the hearts of his countrymen.

Tarasion, who, in spite of an artificial leg, flies his plane so skilfully and with such valour that he brings down seven Boche in two months. Fonk, one of the best, who has more than a score of victories to his credit and to whose lot fell the privilege of avenging the death of Guynemer.

Flaischere, now in America, (at time of first publication), a *pilote extraordinaire*, whose favourite trick is so to outfly and outmanoeuvre his adversary that he forces him to land behind our lines without firing a shot at him, capturing the pilot and the machine intact.

The list might be prolonged indefinitely. Let it not be thought by any one that the mention of these few names implies any invidious comparison. These men are but types, examples given to illustrate the kind of men composing the Allied air forces and the spirit that pervades the whole service. God grant that their kind may flourish and increase and may He give them, one and all, added courage, strength and skill to the end that they may finally issue forth from the battle for Liberty, Democracy and Peace bearing on high and unfurling to the four winds of earth the banner of HONOUR AND VICTORY.

CHAPTER 8

America in the Air

France has been very generous in her recognition of the efforts and services of the Americans who have devoted their strength or skill, and dedicated their lives, to her cause in the various departments of her army. It is her custom to recognize special or exceptional acts of bravery, valour or sacrifice, on the part of any of her warriors, whether they be French or men of other nationalities who have volunteered to serve her, and to publish her appreciation to the world by means of citations.

These citations may originate from different sources and the degree of honour implied is somewhat in accordance with the importance of the source. First, in importance and in the honour conferred, is the *Corps d'Armée* citation, which is issued by the general commanding the Army Corps to which the individual who is to be honoured is attached. Next, comes the Divisional citation, issued by the general commanding a Division. Lastly, there is the Regimental citation, issued by the colonel commanding a Regiment,

A man's first citation, without regard to which of the three sources named is responsible for it, always carries with it the *Croix de Guerre* as a personal decoration and the perpetual right to wear it as a badge of honour. The *Croix de Guerre* is a bronze Maltese Cross suspended from a pin by a red and green ribbon; there are, however, three grades of this decoration, each designated by a distinguishing mark to indicate the relative rank of the officer bestowing it or the degree of honour implied. The first and highest class, that of the *Corps d'Armée*, is designated by a bronze palm leaf attached to the ribbon; the second, or Divisional class, by a gold star, similarly attached; and the third, or Regimental class, by a bronze star. For each new citation which a man receives he is entitled to wear an additional emblem of the class corresponding to the source of the citation, as explained.

The *Médaille Militaire*, with the yellow and green ribbon, is the decoration next higher than the *Croix de Guerre* and is given almost exclusively to men in the ranks for exceptional bravery. It is almost impossible for an officer to win this decoration and it is considered an exceptional honour if it is bestowed upon one.

The next decoration is that of the *Legion d'Honneur*. This is bestowed, almost without exception, upon officers; very few men in the ranks have ever been given the medal of the Legion of Honour. There have been many Americans who have received the *Croix de Guerre* or the *Médaille Militaire*, or both, but only two, so far as I am aware, have been honoured by the Legion of Honour to this date; these two are William Thaw and Raoul Lufberry.

Another decoration, known as the *Fourrageres*, is a three strand braided silk cord, the end tipped with a silver or gold ornament, which the soldier wears over and under his left shoulder. This is given to a whole regiment or squadron for conspicuous bravery or exceptionally meritorious service collectively, or as a unit. It always marks the wearer as a member of a crack regiment or *escadrille*. Every member of an organization upon which it has been bestowed is entitled to wear it, whether he be a regimental officer or private, or a captain, pilot, *mechanician* or cook of an *escadrille*.

There are three grades of the *Fourrageres* which are differentiated by colours of the braid corresponding to the colours of the ribbons on the medals already mentioned; that is, a green and a red braid indicates that the honour bestowed upon the organization is equivalent to the *Croix de Guerre* when given to an individual; green and yellow corresponds to the Military Medal; and solid red to the Legion of Honour.

Out of the hundreds of *escadrilles* in the French Army only about a dozen have been honoured by any grade of the *Fourrageres*; but all three grades have been conferred upon the Foreign Legion, which is the only military organization to be so signally honoured up to the present time.

There is a badge, consisting of a blue and yellow striped bar, known as the *Medaille de Reforme*, given to all those who have been wounded in active service. A plain bar indicates a minor wound, which may receive only first aid attention or be dressed at the front; the same bar with a red star attached means that the wearer has received a wound serious enough to warrant removing him from the front to a base hospital for treatment.

The more prominent living Americans in the French air service, and an example of the citations given to each one, follow:

William Thaw, Lieutenant, *Escadrille N-124.* An excellent pilot who has always been an example of spirit and courage. He returned to the front after recovering from a serious injury. During the German retreat he gave proof of his initiative and intelligence by landing near the troops on march and communicating information concerning the enemy's movements, gained while flying at a low altitude, as the result of which surprises were avoided. He brought down an enemy plane on April 26th, 1917.

Raoul Lufberry, Adjutant Pilot, *Escadrille N-124*, has been named for the Order of the Legion of Honour, with rank of *Chevalier.* He enlisted under the French flag for the duration of the war. He has given proof of remarkable daring as a pursuit pilot and, up to December 27th, 1916, brought down six enemy planes. Has already been cited twice in the orders of the army and has been given the Military Medal.

(A second citation). A marvellous pursuit pilot. He is a living example to his *escadrille* of boldness, coolness and devotion. On the 10th of June, 1917, he brought down his tenth enemy plane.

Didier Masson, Adjutant Pilot, *Escadrille N-124*, is given the Military Medal, After taking part in a number of artillery regulations and reconnaissances, he bravely participated in the operation of a group of pursuit planes at Verdun. On the 12th of October, 1916, while protecting a bombing plane, he brought down an enemy machine. He accomplished this mission in spite of a leak in his gasoline tank, occurring over the German lines, obliging him to plane back to our lines without power. This citation carries with it the Croix de Guerre with palm.

Charles Chouteau Johnson, Sergeant, *Escadrille N-124.* An American citizen, enlisted for the duration of the war. A good pilot, who has rendered excellent service to his *escadrille* at Verdun and on the Somme. On the 26th of April, 1917, he attacked and brought down an enemy plane.

Willis Haviland, Sergeant, *Escadrille N-124.* An American citizen, enlisted for the duration of the war. A good pilot, courageous and skilful. Attacked an enemy plane on the 26th of April, 1917, and brought it down inside the first German lines.

Horace Clyde Balsey, Corporal, *Escadrille N-124*. The Military Medal has been conferred on this soldier, who enlisted for the duration of the war, A young pilot, full of action and courage. On the 18th of May, 1916, he attacked several enemy combat planes behind their own lines. Seriously wounded in the course of the combat, he managed to bring his machine back to the French lines. The present citation carries with it the *Croix de Guerre* with palm.

James Norman Hall, Corporal, *Escadrille N-124*, an American volunteer, while flying over the enemy lines on June 16th, 1917, gave battle to seven hostile planes and in spite of the fact that he sustained three wounds, at least one of which was very serious, and temporarily lost consciousness, he succeeded in landing his machine within our first line trenches. He has been decorated with the *Croix de Guerre* with palm, and with the *Médaille Militaire*.

Frederick Zinn, Observer, *Escadrille F-24*, an American volunteer, enlisted in the Second Foreign Legion; participating in all of the operations of this organization from August, 1914, to October, 1915. After being seriously wounded, he was transferred to the aviation service as an observer, where he immediately distinguished himself by his coolness, courage and disregard of danger. Since the 10th of April he has taken, often without protection, a large number of aerial photographs, which have always been well done, in spite of hostile artillery fire and the attacks of enemy planes.

William Dugan, soldier of the 1st Company of the 170th Regiment of Infantry. In the attack on the 1st of May, 1916, he showed great bravery in the assault on the enemy trenches and took several prisoners. [Some time after this citation Dugan became a member of the Lafayette Escadrille.]

The men whom I have just named are all living at the time this is written. But let us turn a moment to do simple homage to some of those who have passed on to join that band of Spirit Knights who, let us believe, soar above these living Knights of the Air and watch the constant struggle that is being waged below them. Much has been written by better pens than mine of Victor Chapman, Kiffen Rockwell, Norman Prince, and James McConnell, and I shall not attempt to recount their exploits here; but I mention their names, in passing, as testimony to their courage and valour, and in simple tribute to their memory.

But of some who have fallen more recently very little, so far as I know, has been written, and, therefore, I shall attempt to speak of some of these, even though it be briefly and insufficiently.

Edmund Charles Genêt, in the early part of 1915, at the age of eighteen, was in the Foreign Legion, the comrade of William Thaw, Kiffen Rockwell, and Norman Prince. He remained in the Legion for some time after his three friends entered the aviation service, taking part in the big offensive in the Champagne from September 24th to 28th, 1915, when he was wounded in action. Sometime later he followed his three comrades into the aviation service and eventually joined their *escadrille*. He was James McConnell's flying partner on the day that McConnell was brought down by the enemy, March 19th, 1917, and a little less than a month later, on April 16th, 1917, in the region of Ham, he too met his death at the hands of the Hun.

Shortly after this Ronald Wood Haskier fell, and then Lief Norman Barclay. Barclay had the rare distinction of being cited in the orders of the day even before he had arrived at the front for active service. In testing out a plane one day his flying wires broke, throwing him into a spinning nose dive with one wing gone. In spite of his perilous predicament, calculated to try the nerve of the most experienced pilot, he preserved his presence of mind and succeeded in regaining control of his machine, difficult as it was on account of the broken wing, and made his landing safely. His citation praised his coolness, his persistence, and his determination in the face of danger, a remarkable testimony to his ability. But he was doomed to fall in combat about two months later.

James Oliver Chadwick, of *Escadrille N-73*, was named by the chief instructor of the Bleriot school as the best American and most' consistent pilot who had come under his observation, one who applied brains to his flying) both in theory and practice, as he had in the study and practice of law. He was one of the most popular members of the aviation corps; a magnificent athlete, six feet two inches in height, and an upstanding young American of the best type. He was a graduate of the Harvard Law School, methodical in his training and flying, and thorough in all things.

On August 14th, 1917, Chadwick went out on a patrol with Adjutant Paris, his flying companion. After reaching a considerable height, Paris saw a German plane directly over Chadwick's machine, diving and shooting down at him. He just had time to see Chadwick make a quick *renversement* in order to get behind the Boche when he, Paris, was compelled to make a similar manoeuvre. When Paris recovered from his *renversement* and looked for Chadwick, both he and the Boche had disappeared. Paris scouted around for some time, endeavouring to catch sight of his mate, and then returned to the aerodrome, hoping that Chadwick had preceded him. But Chadwick was never seen again by any of our men. Six days later, after a quick rush by which our troops gained a little of the enemy's ground, Chadwick's machine, a Spad, was found—the Boche had not had time to salvage it—but no trace of our friend. We may never know how his end came, but it is certain that he met it, as he met all things in life, like a true American gentleman.

Then in August, 1917, Julian Biddle, sterling and intrepid American, scion of an old and honoured family, met his fate, falling into the English Channel and being drowned.

Charles Trinkard, *Ancien Légionnaire*, was the next to go. "Tiny Trink" was Trinkard's favourite nickname among the boys. He stood but an inch over five feet in height, but he was every inch a soldier and every pound a fighter. When the war broke out he was a time-keeper, working for the *Compagnie Generale Transatlantique* (French Line), on Pier 57, in New York city. Twenty-one days later he was in Paris, enlisted in the French army. From August 24th, 1914, he was in the famous *Premier Regiment* of the Foreign Legion; he fought in the big attacks in Champagne, where he was wounded in September, 1915; later he rejoined his regiment and participated in the Somme attack and was again wounded. During part of this time he was the comrade and chum of Alan Seeger, whose remarkable poems have received such favourable notice. In the earlier days of the war he served with Chapman, Rockwell, Thaw, Bach, Dowd, Bouligny, Zinn, Hall, Chatkoff, Soubrian, and Scanlon. More than once, on account of his height, his comrades were obliged to give him a boost from the trenches to enable him to climb "over the top."

In February, 1917, Trinkard was transferred to the aviation service. He was so small he could hardly reach the rudder bar of the plane with his feet, but the same clean grit and tenacious spirit carried him along and in due time he was a breveted pilot. However, before he received his brevet, a *penguin*, ran him down and broke his arm. This unfortunate experience was a standing joke among the other aviators and, in spite of the fact that everybody loved and respected him, he was never able to escape this jest.

In September, 1917, after three long years of war in all of its ugliest phases, while on flying duty at the front near Nancy, he felt the call of home. If ever a man led a charmed life that man appeared to be Trinkard, after all the dangers he had survived. But he seemed to feel that the charm was beginning to lose its potency and he applied for a three weeks' leave in America. On November 26th, 1917, he fell to his death and *one hour later* his permission to return to America arrived at his aerodrome. But dear little "Trink" did not need the permission then, he had gone on a longer journey.

There are others who have passed on, some quite as gloriously as those mentioned, but the limits of space forbid details concerning them here. However, I must mention two who are living, perhaps, the life of death, prisoners in the hands of the enemy.

James J. Bach, once a member of the Foreign Legion, and one of the nine men who formed the first American Escadrille, volunteered, during the summer of 1915, to accompany a French pilot, each flying a two-seater plane, on an important and perilous mission behind the German lines. They were ordered to pick up some French spies at a certain point and bring them back to our lines. Reaching the designated spot, the French pilot landed unobserved, but in landing he crashed his plane, which then burst into flames. Bach, hovering a thousand feet above, saw the accident and also a Boche patrol of cavalry, which had been attracted toward the spot by the flames, and he immediately dived for the field to rescue his comrade. But, unfortunately, when he landed his motor stalled and before he could start it again and "take off" with the Frenchman the patrol pounced upon them and both men were taken prisoners. He was accused of being a spy and was tried twice in Berlin by the Council of War on that charge, but was finally acquitted. However, he was held, and still remains, a prisoner in Germany, (at time of first publication).

Everett Buckley, an Illinois boy, fell with his plane in flames over Dixmude, in Belgium, on September 6th, 1917. He was mourned in France and at home as dead, but during the following November a letter from a prison hospital in Germany was received from him stating that he was recovering from his burns and was safe and well otherwise. But he, too, must probably spend his days during the remainder of the war in the clutches of the Hun.

It is perhaps fitting that something should be said briefly of the conception and organization of that body of American fliers which was known as the American Flying Corps.

The idea originated with two men at Marblehead, Massachusetts, in December, 1914, and to them should be given the credit for formulating the plan which was finally carried through to a successful organization. These two men were Frazier Curtis and Norman Prince, both of whom were sea-plane pilots at that time. They succeeded in interesting a number of other men whose sympathies were already strongly with France, though at that time the war was only fairly well begun and public sentiment in America had not yet had time to become thoroughly crystallized.

But these men were blessed with a vision by which they saw an opportunity to do a real service for France and for Liberty and at the same time repay a part of the debt which America owes to Lafayette, Rochambeau and other men of France.

Early in 1915 these young men arrived in France and, together with seven others, offered themselves to the French Air Service. But it was not until February 20th that they were finally accepted and inducted into the French service. The nine men who were the original members of the American Flying Corps, and who began their training as aviators for France on that date, were: James J. Bach, Elliot Cowdin, Frazier Curtis, H. G. Gerin, Bert Hall, D. D. Masson, Norman Prince, Andrew Ruel, and William Thaw.

They began their training at Pau, and later on were sent to Le Bourget; from there they went to the front and were assigned to duty in Voisin bombing planes. This was the nucleus of what was then known as the American Flying Corps, or, as it was sometimes called, *The American Escadrille*. It was in the early part of June, 1915, that they began fighting the Germans, and they have continued to do so up to the present time.

However, because of an agitation started in America, undoubtedly of pro-German origin, by which objection was registered against the use of the name, American Flying Corps, on the ground that it was a violation of America's neutrality at that time, legislation was secured which forbade the use of that name. This legislation went further than this, in so much as it caused all Americans serving in the armies of a foreign country at that time to lose their status as United States citizens. Therefore, the name of the American Flying Corps was changed to the *Franco-American Flying Corps*, and the title of the squadron was changed to the *Lafayette Escadrille*. [1]

By the time January, 1917, had arrived the Lafayette Escadrille had won such renown, and its name and fame were so widely known, that it was decided to change the name from the Franco-American Flying Corps to the *Lafayette Flying Corps*, a name now known the world around.

The reader should understand that there is a difference between a Flying Corps and an Escadrille. The former includes the latter. The *escadrille* is really a squadron of aviators, usually consisting of fifteen men, together with the necessary *mechanicians* and other helpers, who are on active duty *at the front* and who are constantly flying over the lines of the enemy. The Flying Corps embraces all of the men included within a certain classification (all American fliers, in this instance) whether attached to any of the numerous French *Escadrilles* or in training at one of the schools in the rear.

Of the men who have been attached to the Lafayette Flying Corps, the most famous and successful is the great American Ace, Raoul Lufberry, recently Second Lieutenant in the French Aviation Service; now Major in the Aviation Section of the United States Army, (at time of first publication). Much has been written concerning him and much more could be said, for his career since early youth reads like romance. Leaving his home in Wallingford, Connecticut, at the age of seventeen, he set forth to see the world.

As he was of French ancestry, France was the first country to attract him. Then he visited Egypt and other points in Africa, Turkey, Germany, and eventually South America. Returning home he enlisted in the American Army and was sent to the Philippines. Completing his service there, he roamed through China, Japan, and India, and finally happened to visit Saigon, in Cochin China, at the time Marc Pourpe, the intrepid French aviator, was giving exhibition flights in Asia. It was then and there that Lufberry first felt the call and fascination of aviation. He met Pourpe, applied to him for a position as *mechanician*, and was employed. He remained in Pourpe's service for a long time, travelling with him from place to place, and early in 1914 helped to tune and nurse the motor that enabled Pourpe to make his famous flight from Cairo to Khartoum.

When the war broke out Pourpe offered his services as an aviator to France, and Lufberry went with him. Pourpe was killed in action in December, 1914, and then Lufbery, who had continued to act as

1. *The Story of the Lafayette Escadrille* by George Thenau also published by Leonaur.

Pourpe's *mechanician* up to that time, determined to be the personal avenger of his friend and patron. And what an avenger he has been! His present record is, no doubt, far beyond his own dreams at that time. He succeeded in having his status in the French service changed from *mechanician* to that of *Elève Pilote* (Student Pilot) and began his training as an aviator, which was finished in the early summer of 1916. He began active service at once and brought down his first enemy plane on July 30th, 1916; another followed on August 4th—two victories within less than a month's service. For this he was decorated with the *Médaille Militaire*, and a palm was added to his *Croix de Guerre*, which had already been given to him. His citation, issued at that time, read:

> Raoul Lufbery, Sergeant in *Escadrille N-124*. A model of address, of coolness and of courage. He has distinguished himself in numerous long distance bombardments and by several combats in which he put enemy machines to flight. On July 80th he was attacked at close range by a group of four German planes and brought down one of them within our lines. On August 4th, 1916, he brought down his second machine.

He continued his excellent work, bringing down another Boche four days after that. Then during the bombardment of the Mauser works, October 12th, 1916, he added two more machines to his credit, making his fifth official victory, and giving him the rank of an Ace; all of this within about three months' time. It was during this bombardment of the Mauser works that Norman Prince was mortally wounded.

So Lufberry has gone on and on, from victory unto victory, seemingly a spirit of vengeance that exacts its toll of a plane for a plane, every time an American lad is sacrificed to the Boche—and perhaps a few extra for good measure. On the Somme, in November, he forced two Boche planes to the earth at one time, but they fell too far behind the enemy lines to be observed officially, so they are not included in his authenticated record. But on December 27th he got a German *Aviatik* back of our lines. This combat nearly cost him his life, as four of his adversary's bullets passed very close to his body, but it did not deter him in the least. A citation in August, 1917, mentions the fact that he engaged in nineteen combats in one day's work of three hours.

Lufberry is a calm, collected sort of chap, he does not take himself too seriously and is inclined to be reticent. His chest is emblazoned with many decorations, as the result of the great work he has done, yet with it all he is modest and unassuming and one of the most popular pilots in France.[2]

William Thaw, at one time the Commanding Officer of the Lafayette Escadrille, is now a Major in the Aviation Section of the United States Army, (as at time of first publication). Thaw, at the time this is written, has not attained the rank of Ace;[3] this title being given to those aviators who have brought down five or more enemy planes within sight of observers who can certify to the facts and thereby give official credit for each one. But when it comes to playing the game, as leader of the patrol and as a cool, intrepid and determined pilot, he is more than an Ace. For it is he who sets the example of courage and fearlessness, tenacity and perseverance, that helps and heartens the younger pilots and inspires them with the will to succeed.

Whenever there is a call for pilots to make a long or hazardous trip beyond the lines, whether it be to raid a Boche aerodrome or to shoot down their observation balloons, it is Thaw who is the first to volunteer and Thaw who leads the others to the work which may be undertaken. Always on the job, every day, with his keen perception of duty; always leading his men, many of whose names you have seen mentioned in the daily papers as the result of their exploits in the French service, and whose names you will see again when the time comes to recount the deeds done by the American Aerial Forces over there.

It is fitting that mention should be made of the patriotic Americans who helped to father the idea of the first American Flying Corps and by whose substantial aid the project was made possible and practical. Therefore, in this connection, I shall name: Mr. William K. Vanderbilt, the Honorary President of the Lafayette Flying Corps; Mons. M. de Sillac, the President, and Dr. Edmund Gros, the Vice-President.

Too much could hardly be said of what our American boys have done, and much may be predicted for what they will yet do when the time is ripe and circumstances warrant their full activity. The list as it stands today is long and each has his individual victories that are worthy of comment and praise; but I cannot attempt to touch upon more than these few, nail sketches of some of the more prominent, may catch a glimpse of the spirit and purpose which these men typify and which shall be an example that others who are to follow may en-

---

2. Major Lufberry's death has been reported recently, (as at time of first publication).
3. Thaw brought down his fifth plane and became an Ace in April, 1918.

deavour to emulate and by which they may measure their own efforts. It is safe to predict that there will be others who will equal the record and high standard already set for them by the men who have carried the French *cocarde* on their planes and who are now ready to carry the American emblem into the air above the enemy's territory.

LEONAUR

ALSO FROM LEONAUR
AVAILABLE IN SOFTCOVER OR HARDCOVER WITH DUST JACKET

WINGED WARFARE *by William A. Bishop*—The Experiences of a Canadian 'Ace' of the R.F.C. During the First World War.

THE STORY OF THE LAFAYETTE ESCADRILLE *by George Thenault*—A famous fighter squadron in the First World War by its commander..

R.F.C.H.Q. *by Maurice Baring*—The command & organisation of the British Air Force during the First World War in Europe.

SIXTY SQUADRON R.A.F. *by A. J. L. Scott*—On the Western Front During the First World War.

THE STRUGGLE IN THE AIR *by Charles C. Turner*—The Air War Over Europe During the First World War.

WITH THE FLYING SQUADRON *by H. Rosher*—Letters of a Pilot of the Royal Naval Air Service During the First World War.

OVER THE WEST FRONT *by "Spin" & "Contact"* —Two Accounts of British Pilots During the First World War in Europe, Short Flights With the Cloud Cavalry by "Spin" and Cavalry of the Clouds by "Contact".

SKYFIGHTERS OF FRANCE *by Henry Farré*—An account of the French War in the Air during the First World War.

THE HIGH ACES *by Laurence la Tourette Driggs*—French, American, British, Italian & Belgian pilots of the First World War 1914-18.

PLANE TALES OF THE SKIES *by Wilfred Theodore Blake*—The experiences of pilots over the Western Front during the Great War.

IN THE CLOUDS ABOVE BAGHDAD *by J. E. Tennant*—Recollections of the R. F. C. in Mesopotamia during the First World War against the Turks.

THE SPIDER WEB *by P. I. X. (Theodore Douglas Hallam)*—Royal Navy Air Service Flying Boat Operations During the First World War by a Flight Commander

EAGLES OVER THE TRENCHES *by James R. McConnell & William B. Perry*—Two First Hand Accounts of the American Escadrille at War in the Air During World War 1-Flying For France: With the American Escadrille at Verdun and Our Pilots in the Air

U-BOAT WAR 1914-1918 *by James B. Connolly/Karl von Schenk*—Two Contrasting Accounts from Both Sides of the Conflict at Sea During the Great War.

LEONAUR

ALSO FROM LEONAUR
AVAILABLE IN SOFTCOVER OR HARDCOVER WITH DUST JACKET

"AMBULANCE 464" ENCORE DES BLESSÉS *by Julien H. Bryan*—The experiences of an American Volunteer with the French Army during the First World War

THE GREAT WAR IN THE MIDDLE EAST: 1 *by W. T. Massey*—The Desert Campaigns & How Jerusalem Was Won---two classic accounts in one volume.

THE GREAT WAR IN THE MIDDLE EAST: 2 *by W. T. Massey*—Allenby's Final Triumph.

SMITH-DORRIEN *by Horace Smith-Dorrien*—Isandlwhana to the Great War.

1914 *by Sir John French*—The Early Campaigns of the Great War by the British Commander.

GRENADIER *by E. R. M. Fryer*—The Recollections of an Officer of the Grenadier Guards throughout the Great War on the Western Front.

BATTLE, CAPTURE & ESCAPE *by George Pearson*—The Experiences of a Canadian Light Infantryman During the Great War.

DIGGERS AT WAR *by R. Hugh Knyvett & G. P. Cuttriss*—"Over There" With the Australians by R. Hugh Knyvett and Over the Top With the Third Australian Division by G. P. Cuttriss. Accounts of Australians During the Great War in the Middle East, at Gallipoli and on the Western Front.

HEAVY FIGHTING BEFORE US *by George Brenton Laurie*—The Letters of an Officer of the Royal Irish Rifles on the Western Front During the Great War.

THE CAMELIERS *by Oliver Hogue*—A Classic Account of the Australians of the Imperial Camel Corps During the First World War in the Middle East.

RED DUST *by Donald Black*—A Classic Account of Australian Light Horsemen in Palestine During the First World War.

THE LEAN, BROWN MEN *by Angus Buchanan*—Experiences in East Africa During the Great War with the 25th Royal Fusiliers—the Legion of Frontiersmen.

THE NIGERIAN REGIMENT IN EAST AFRICA *by W. D. Downes*—On Campaign During the Great War 1916-1918.

THE 'DIE-HARDS' IN SIBERIA *by John Ward*—With the Middlesex Regiment Against the Bolsheviks 1918-19.